Dinosaur Days in Texas

Acknowledgements

The publisher gratefully acknowledges the assistance provided by the Texas Parks and Wildlife Department. Particular thanks to those individuals in the Interpretation and Exhibit Branch whose skills and dedication make a visit to our state parks more enjoyable. Dinosaur State Park is an outstanding example.

We also wish to recognize the help and information provided by Dr. Jim Farlow, Dept. of Earth and Space Sciences, Indiana University-Purdue University at Fort Wayne; Dr. Wann Langston, Jr., Professor Emeritus, Dept. of Geological Sciences, University of Texas at Austin; School of Geosciences, Louisiana State University at Baton Rouge and the American Museum of Natural History, New York City.

Dinosaur Days in Texas

by
Tom and Jane D. Allen
with Savannah Waring Walker

Hendrick-Long Publishing Co.
DALLAS

Library of Congress Cataloging-in-Publication Data

Allen, Tom, 1930-
    Dinosaur days in Texas / by Tom and Jane D. Allen with Savannah Waring Walker.
        p.   cm.
    Bibliography: p.
    Includes index.
    Summary: Discusses the various species of dinosaurs discovered in Texas and where their fossil remains can be seen.
    ISBN 0-937460-30-3
    1. Dinosaurs—Texas—Juvenile literature. [1. Dinosaurs—Texas. 2. Paleontology—Texas.] I. Allen, Jane D. II. Walker, Savannah Waring. III. Title.
QE862.D5A39 1989   88-37237
567.9'1'09764—dc19   CIP
                                                                       AC

© 1989 Hendrick-Long Publishing Co.

Hendrick-Long Publishing Co., Dallas, Texas
All rights reserved.
Printed and bound in the United States of America.

Illustration Credits

Cover: Nola Montgomery, Texas Parks and Wildlife Dept./ **10:** Nola Montgomery, Texas Parks and Wildlife Dept./ **11:** Flying Colors Studio/ **13:** Wann Langston Jr./ **14: 17: 19: 21: 22: 23: 25: 26: 27:** Flying Colors Studio/ **28: 29: 30: 31:** dinosaurs, Aaron Morris — Texas Parks and Wildlife Dept.; charts, Flying Colors Studio/ **32:** fossil trackways, School of Geosciences - Louisiana State University, Baton Rouge; dinosaur herd, Nola Montgomery, Texas Parks and Wildlife Dept./ **33: 34: 35: 36: 37: 38: 39: 40: 41: 42: 43:** dinosaurs, Aaron Morris — Texas Parks and Wildlife Dept.; charts; Flying Colors Studio/ **44:** Pteranodon, Bureau of Economic Geology, University of Texas, Austin; size comparisons, Flying Colors Studio/ **45:** Rhamphorhynchus, Bureau of Economic Geology, University of Texas, Austin; Quetzalcoatlus, Aaron Morris/ **46:** Mosasaur, Aaron Morris; skeleton, Dallas Museum of Natural History/ **47:** Phobosuchus, Aaron Morris/ **48:** Protostega, Aaron Morris/ **49:** map, Texas Parks and Wildlife Dept./ **53:** Nola Montgomery, Texas Parks and Wildlife Dept./ **54: 56:** Flying Colors Studio.

Photo Credits

12: Dr. Jim Farlow, Indiana Univ. - Purdue Univ., Ft. Wayne/ **47:** Neg. Trans. No. 318651 — Photo by C. H. Coles, Courtesy Dept. of Library Services, American Museum of Natural History/ **48:** Strecker Museum, Baylor University, Waco/ **49:** Neg. Trans. No. 319835 — Photo by R. T. Bird, Courtesy Dept. of Library Services, American Museum of Natural History/ **50:** Panhandle Plains Historical Museum - Canyon, Texas/ **54:** Dr. Jim Farlow, Indiana Univ. - Purdue Univ., Ft. Wayne/ **55:** Dr. Jim Farlow, Indiana Univ. - Purdue Univ., Ft. Wayne/ **57:** Ft. Worth Museum of Science and History.

Book design by Flying Colors Studio
Type by Express Typesetting Company

Table of Contents

Part One In the Age of the Dinosaurs 11

Part Two Dinosaurs in Texas 27

Coelophysis	28	Ornithomimus	39
Technosaurus	29	Panoplosaurus	40
Acrocanthosaurus	30	Stegoceras	41
Deinonychus	31	Torosaurus	42
Pleurocoelus	33	Tyrannosaurus	43
Iguanodon	34	Pteranodon	44
Tenontosaurus	35	Rhamphorhynchus	45
Alamosaurus	36	Quetzalcoatlus	45
Chasmosaurus	37	Mosasaur	46
Edmontosaurus	38	Phobosuchus	47
Hadrosaurus	38	Protostega	48

Part Three The State's Dinosaur Attractions . . . 49

Maps/Charts

Fossil Finds	14
Earth Clock	17
Geologic Time Scale	19
The Ancient Continents	21
Dinosaurs in Texas	25
Locations of Dinosaur Tracks	26
Glossary	57
Bibliography	61
Index	63

Foreword

There are always strange stories about unusual creatures that are too big, too strange, and too powerful to believe in. Does a huge beast swim secretly in the murky waters of Loch Ness in Scotland? Does a hairy ape-man walk the icy snow fields of the Himalayan Mountains? We hear the stories, but as time goes by and proof is scarce, we begin to doubt that anything so unusual really exists. We begin to suspect that "what we see" in this world is "what we get."

There was a time, not far back in human history, when strange shapes found in the rocks were thrown aside as accidents of nature or "tricks of the devil." Only a few very smart persons realized that nature was trying to tell us a fascinating story about itself. Slowly, scientists began to see that the strange shapes were a record of the past, preserved in stone. Some of those shapes told a story more unbelievable than we could imagine. We came to learn of dinosaurs!

Who would have imagined that what seems like an eternity ago - 65 million years - giant reptiles, bigger than any living animal, walked the earth? Some seemed like monsters from our worst nightmares—toothy beasts with very bad tempers! Suddenly we knew that life had hidden from us a great secret, until we were ready to learn.

Dinosaurs still seem unbelievable, even after we have known about them for over 100 years. We have come to appreciate their many different shapes and their power. We perhaps feel akin to them because, like us, they ruled the world.

At the same time that the secret of the dinosaurs' existence was made known to us, we became aware also that a greater mystery was unexplained. What had caused the disappearance of such big and powerful creatures? Were there powers in the universe more powerful than those of the great dinosaurs—greater than ours?

The dinosaur's story puts a tingle down our spine. There may be other secrets about the past that we need to know. That is why we are so interested in dinosaurs.

This book, *Dinosaur Days in Texas*, has a surprise for us. Dinosaurs lived in Texas! Not only did they live here, but many different kinds of them thrived. In fact, either your favorite dinosaur or something like it lived in Texas. When we read books published in other parts of the United States, we get the feeling that all the interesting dinosaurs must have lived somewhere else. Texas dinosaurs have not been studied as much as some other dinosaurs. The great dinosaur hunters of the last century largely passed Texas by and went to find bones in the rich Jurassic-age rock beds of the far American West. Texas rocks and Texas dinosaur bones had to wait. Just in the last few years have paleontologists (the scientists who study fossils) made big dinosaur discoveries in Texas. That is wonderful, because it means that someone reading this book may be the one to make a great Texas dinosaur discovery of the future!

Many actual dinosaur footprints have been found in the rocks near Glen Rose, Texas—and on many Texas ranches. The number of discoveries of dinosaur fossil bone is now a little over 100, several of those discoveries being nearly complete skeletons. Many of the finds, however, are very small bits of bone that only a paleontologist would know for sure belonged to dinosaurs. So Texas still needs future dinosaur scientists. The greatest discovery is still buried somewhere out there in the Texas hills and prairies.

You will find this book fascinating when you discover how much we know about Texas dinosaurs now. By reading it, perhaps you will want to look for fossil bones—maybe even dinosaurs. Texas rock is rich in such remains and traces of past life. If you do make a discovery, remember to tell a paleontologist about it. Paleontologists work at universities and museums. They are friendly people and are always glad to talk with you about your discoveries. Wouldn't it be fun to have your picture in the newspaper, riding atop your newly found Texas dinosaur bone! Good reading and good dinosaur hunting!

Charles E. Finsley
Curator of Earth Sciences
Dallas Museum of Natural History

Part One

In the Age of the Dinosaurs

Picture this: a giant green dinosaur, three stories high and as long as a house, plods across a muddy plain. Its huge feet make squishy, round tracks. The animal takes enormous steps; each of its tracks is yards away from the next. Alongside this monster runs a smaller one, making vicious, three-clawed prints closer together in the mud.

All of a sudden the smaller dinosaur lunges forward, opening its mouth wide and sinking its razor-edge teeth into the tail of its giant prey. A battle for life has begun.

You don't need to go to the movies to see this story taking place. It isn't made up; it happened in real life, millions and millions of years before people were even here. And you can go to the place where it happened and see the dinosaurs' footprints that prove it.

Dinosaur remains can be seen in many parts of the world. But these particular footprints, in all their living glory, are to be found near the town of Glen Rose, Texas. Glen Rose is just one of the places in this huge state that bear the traces of dinosaurs and the fantastic ancient reptiles that were their kin. In fact, the preserved remains of these great beasts have been dug up in far-flung parts of the state. From Big Bend National Park in the west to the ancient rock formations in the central part of the state, there is proof that many dinosaurs once lived and died in Texas.

Dinosaur trackways in bed of the Paluxy River near Glen Rose. These are the tracks that were excavated in 1940 and sent to several museums. Photograph was taken by R. T. Bird just prior to removal.

This proof that dinosaurs lived is found not only in their preserved footprints. It is also found in their preserved bones and in the imprints that their bones have made in rock surfaces over time. We call all these ancient remains fossils: preserved animal bones; images made in rock by ancient plants and the bodies of ancient animals; and traces of ancient animals, such as footprints or droppings. For a fossil to form, conditions have to be just right. First, a footprint is made—or an animal or plant is trapped—in mud or clay, often underwater. Layers of mud and other matter then begin collecting on top of the trapped specimen. Over the course of millions of years, the mud around the animal, plant, or footprint is pressed down by all the extra layers of matter forming above it. The mud is also pressed by the slow shifting of the earth's surface, as mountains and oceans take shape. All this pressing turns the mud into rock. Bones trapped in the mud also undergo a chemical change as a result of the pressure. What we find as a fossil in the rock is either preserved, chemically changed bone, or a beautiful image of bone, plant matter, or an animal's foot.

People first began finding dinosaur fossils many hundreds of years ago, way before they had any idea exactly what the fossils were. Some thought the bones they had found might have belonged to "dragons," though dragons only exist in fairy tales. Others thought they had found the footprints of ostrich- or elephant-like animals. Eventually, though, scientists began connecting and joining up the fossil bones they had found buried near each other. The fantastic skeletons that these bones made looked like they had belonged to gigantic, lizard-like animals. By the 1840's scientists had come to an important conclusion:

R. T. Bird's reconstruction of the event that produced part of the Glen Rose trackway, as told on page 11.

Fossil Finds — Mesozoic Era

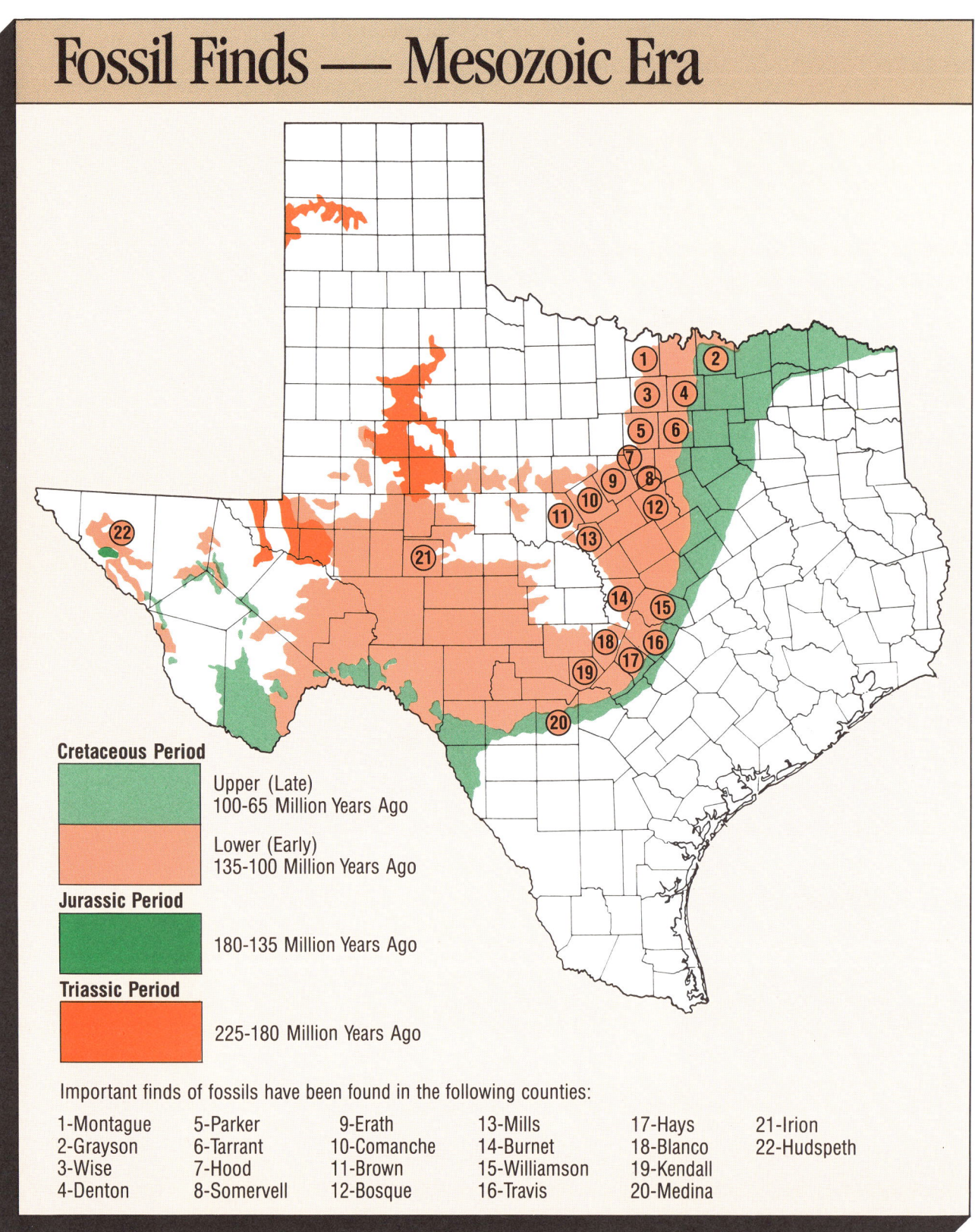

These ancient "lizards" were so different from any animals now living that they should have a name all their own.

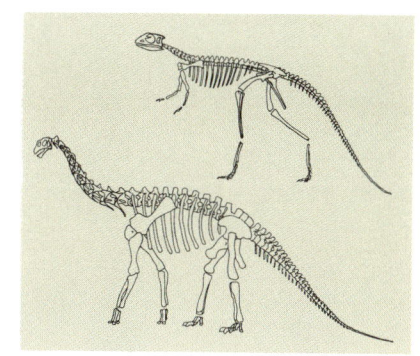

This was when the word *dinosaur* was born. It comes from Greek words meaning "terrible reptile" or "lizard." Each different dinosaur was given a different scientific name—which would often describe the beast's appearance. *Tyrannosaurus*, for instance, was the name given to a particularly vicious-looking, meat-eating dinosaur. The first part of the name, *tyranno-*, comes from a Greek word meaning "absolute ruler," or "tyrant." The second part, *saurus*, is from the Greek word for "reptile" or "lizard."

In spite of these advances, however, no one knew how old dinosaurs were. Scientists had assembled some scary-looking skeletons and named the skeletons. But that was about as far as their knowledge went. It was not until the early part of this century that we came up with a way to test just how ancient these dinosaur fossils really were. This test is called radiometric dating. It is based on scientists' understanding that rocks contain traces of radioactive elements. These elements take a very long time to decay, and scientists have calculated the rate at which this decay takes place. In radiometric dating, the amount of radioactive material that remains undecayed in a particular rock tells the scientist how old the rock is. And however old the rock is, the fossil that was in it is at least that old.

One reason it took scientists so long to develop this test was that their fossil discoveries were giving them second thoughts about a lot of things. They had to get used to the idea that the earth was a lot older than they had thought. They had to understand that each time period in the past brought with it a different layer of rock in the earth's surface—and a different bunch of fossils. They also had to accept that many animals—dinosaurs included—lived and died on earth before we humans ever got here. In fact, much of what scientists were discovering about our world was so different and so new that it took quite a while to sink in.

The scientists' amazement is easy to understand. You, too, might be amazed to know that dinosaurs lived on earth for more than 140 million years. To get some idea of how long a time that is, think about another fact: We humans and our ancestors have only lived on earth 3 million years. In fact, dinosaurs have been gone from the earth for 65 million years. That's 62 million years longer than we have been here!

It is also pretty amazing that scientists have been able to learn all this, and a whole lot more, from studying something that most people ignore; something plain and dull; something that's not alive and never has been: rock. The study of rock has revealed the fossilized animals of bygone ages. The order in which rock layers have been laid down on the earth's surface has told us which of these fossils are older and which are younger. The study of how these rock layers were laid down in the first place has revealed how the earth's continents were formed. And the radiometric testing of rock, as we discussed earlier, has helped us say when all these developments on our planet took place.

You can see from the "Earth Clock" on page 17 that we humans are mere "infants" when you compare our age with that of our planet. If the whole of the earth's visible life were thought of as happening in a

Earth Clock

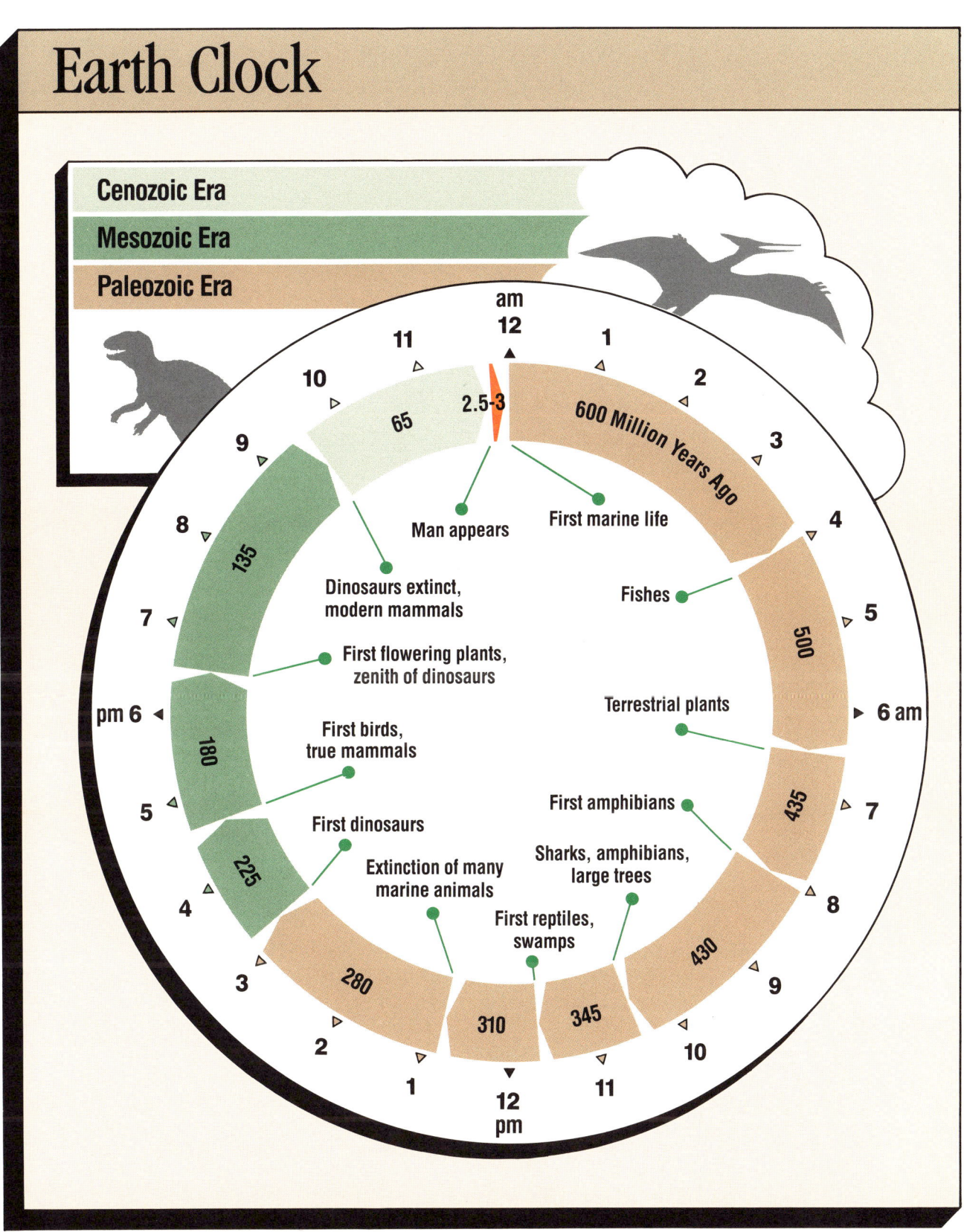

single twenty-four hour day—then humans would only be part of that life for a few minutes at the end of that day. But even so, for "infants," we've done a pretty good job of understanding and learning what has gone before us.

To start with, we know that dinosaurs made their first appearance on earth more than 200 million years ago. At that time the world was made up of a vast amount of ocean and one giant continent, Pangaea. Under the surface of Pangaea, forces were at work that would eventually cause it to break up—to create the oceans, seas and the seven continents we have today. These forces could be seen in earthquakes and volcanoes, among other things. But the effects of the shaking and erupting were too slow to be noticed. And besides, there were no people on Pangaea to notice them anyway.

But even if there had been people, it is important to remember that people might not have noticed the changes either. Because the same forces that broke Pangaea apart are still at work today—right at this moment—slowly changing the shape of the world we live in. The forces are pushing mountain ranges up higher, causing continents to break apart, and drowning certain areas with volcanic lava. But do we notice the changes? Not really. Not unless our own town is being flooded, or we see television pictures of a volcano erupting. And even then we don't think about what our continent will look like in 200 million years—or what it looked like 200 million years ago. We leave those thoughts to the scientists. (Or perhaps we think about becoming scientists ourselves!)

There's no question that scientists have learned a lot about the past (and gotten ideas about the future)

Geologic Time Scale

Cenozoic Era 65 million years ago to present
Mammals, birds, insects and flowering plants abound, with changes more typical of todays animals and plants. Early man appears though not until 2-3 million years ago — long after the dinosaurs became extinct!

Mesozoic Era 225-65 million years ago
The Triassic, Jurassic and Cretaceous Periods comprise the Mesozoic Era. During this time a warm, moist climate supported the growth of palms, ferns, conifers and other plants. A variety of dinosaurs ruled the era. Birds also appeared in the mid-Mesozoic Era.

Cretaceous Period
135-65 million years ago

Jurassic Period
180-135 million years ago

Triassic Period
225-180 million years ago

Paleozoic Era 600-225 million years ago
Marks the appearance of most of the major groups of animals and plants that we recognize today, such as shellfish, insects, spiders, fish, amphibians, reptiles, and most plant types except for the flowering plants.

Note: Geologic time scales are read from the bottom to the top, with the top dates being the most recent dates on the scale.

The Decline of the Dinosaur
A cooler climate during the Cretaceous Period encouraged the development of varieties of flowering plants and grasses. The close of this period is known as "the time of the great dying," during which dinosaurs and other giant reptiles vanished.

The Dynasty of the Dinosaur
During the Jurassic Period the earth's climate continued warm and humid. Dinosaurs dominated the animal world and some reptiles developed the ability to fly.

The Dawn of the Dinosaur
The mild, semi-arid climate of North America grew more humid during the Triassic Period. Reptiles adapted to both land and water habitats, while shellfish abounded in the seas. By the close of the period, dinosaurs outnumbered all other reptiles.

from their study of fossils and rock and earthly forces. As you can see from the chart on page 19 scientists have discovered enough about what happened on earth long ago to be able to name the various times in our planet's ancient history. They can state pretty certainly, for instance, that a world suitable for the dinosaur began 225 million years ago and ended 65 million years ago. They have named that time the Mesozoic Era. The three parts of the Mesozoic Era—the Triassic, Jurassic, and Cretaceous periods—saw three differing groups of dinosaurs, so those names are important to keep in mind as well. During these three periods, dinosaurs roamed the entire mega-continent of Pangaea. Scientists know this because they have found dinosaur fossils on every one of our continents (except Antarctica, where there is too much snow and ice to dig for them). And since our continents were all once a part of Pangaea, it is clear that the great beasts lived just about everywhere.

By the end of the Mesozoic Era, the dinosaurs had all died. Scientists know this because no dinosaur fossils have ever been found in rock formed after the Mesozoic Era. What scientists do not know for sure is why the animals disappeared, although they have come up with many different possible reasons. We will discuss some of the reasons at the end of this book. It is known, however, that the last of the dinosaurs became extinct not too long after the final break-up of the super-continent Pangaea. Pangaea had split far apart over the centuries and centuries that dinosaurs had lived on it. In fact, even by the end of the Triassic Period—the first part of the Mesozoic Era—Pangaea had already split far enough apart to form two big continents, Laurasia in the north and Gondwanaland

The Ancient Continents

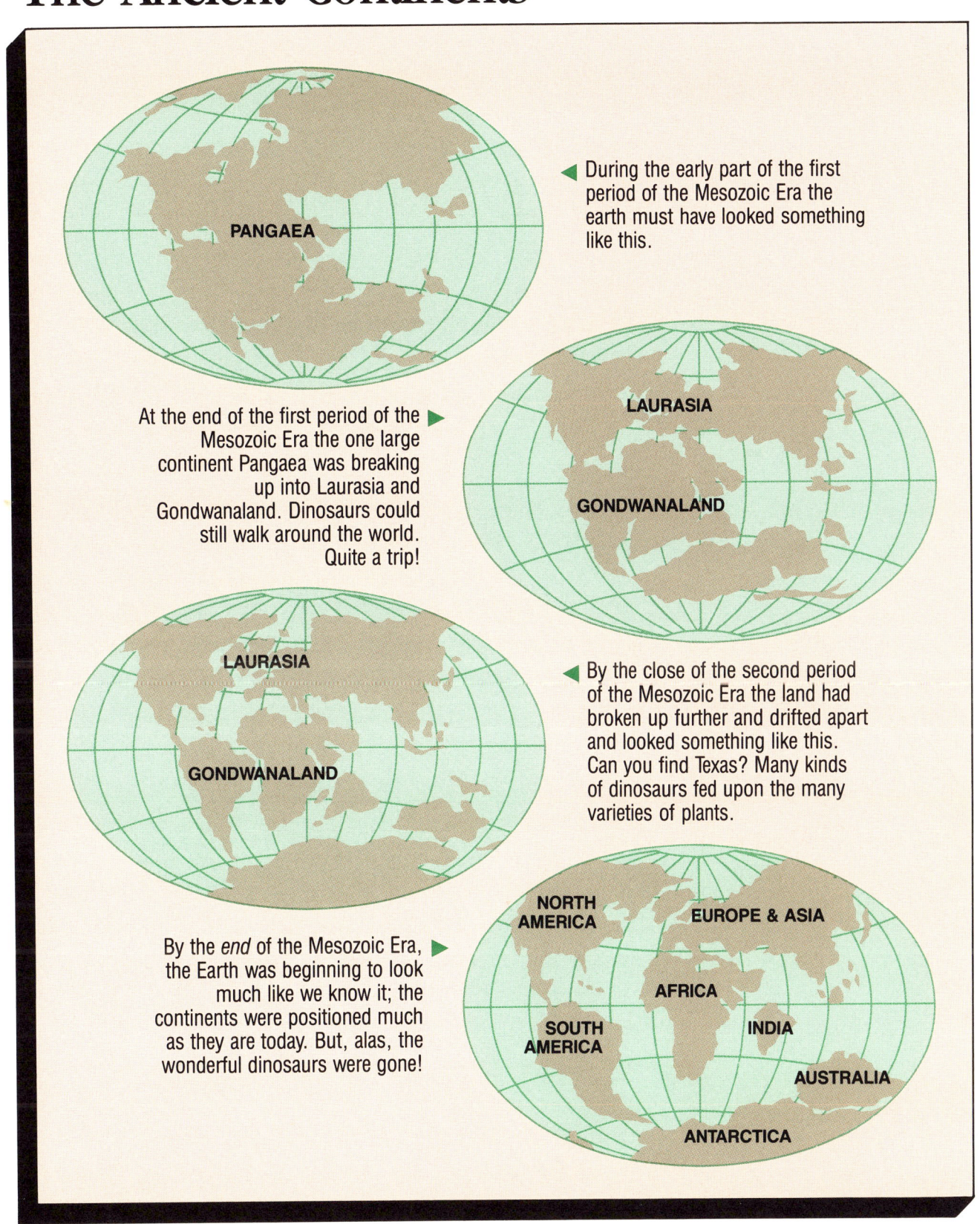

During the early part of the first period of the Mesozoic Era the earth must have looked something like this.

At the end of the first period of the Mesozoic Era the one large continent Pangaea was breaking up into Laurasia and Gondwanaland. Dinosaurs could still walk around the world. Quite a trip!

By the close of the second period of the Mesozoic Era the land had broken up further and drifted apart and looked something like this. Can you find Texas? Many kinds of dinosaurs fed upon the many varieties of plants.

By the *end* of the Mesozoic Era, the Earth was beginning to look much like we know it; the continents were positioned much as they are today. But, alas, the wonderful dinosaurs were gone!

to the south. At that time the two new continents were still fairly close together, but during the rest of the Mesozoic Era water gradually filled the areas between them. The water deepened and widened until it could not be crossed by dinosaurs. Pangaea's split into two and then several continents was not the cause of the dinosaurs' death, but it kept some of the beasts from moving to find warmer climates and more plentiful food as they needed to. So the end of the Mesozoic Era brought with it two other endings: the end of the earth's super-continent and the end of the dinosaurs.

Now you have some idea about how long the dinosaurs lived and where they lived. What you need to know next is a little about the dinosaurs' relation to one another and to other reptiles. These relationships are fairly involved because there were so many different kinds of dinosaurs over the millions of years and they lived over such a vast area. We won't go into any great detail beyond discussing a few of the general groupings of the dinosaurs' "family tree."

To start with, since dinosaurs were reptiles, it would be good to understand their relation to the reptiles that live on earth today. From the study of fossils as well as living reptiles, scientists have learned that our earth has been home to four main groups of reptiles. These groups are different because their skulls are differently formed. The names of the groups all end with the syllable *-sid. -Sid* comes from a Greek word meaning "loop," or "opening."

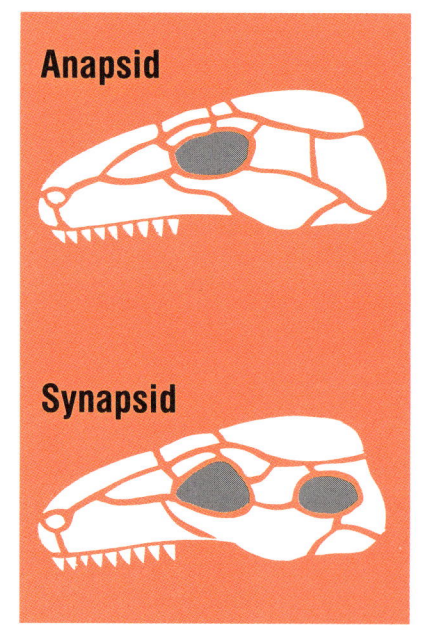

The first group, the *anapsids*, have a solid skull with no openings in it other than the eye sockets. The *anapsids* were a very primitive form of reptile. Most of them lived before the time of the dinosaurs. The only *anapsids* living today are the turtles and tortoises.

The second group is called the *synapsids*. The *synapsid* skull has a single opening (other than the eye sockets). It is found down low on the side of the skull. *Synapsids*, too, lived mainly before the heyday of the dinosaurs. They disappeared for good during the Triassic period. Of all the reptiles, they were the ones who were built the most like mammals. Scientists believe that the world's earliest mammals were descendants of the *synapsids*.

The name of the third group is the *euryapsids*. The *euryapsid* skull also has a single side opening, but it is located a little higher than in the *synapsid* skull. The *euryapsids* all lived in the sea. They vanished at the end of the Mesozoic Era, and they have no living descendants.

The fourth group and the one that is most important to us is the *diapsids*. The *diapsid* skull has two side openings. The dinosaurs that lived on land and glided in the air were all *diapsids*. Ancient crocodiles were too. So are modern-day crocodiles, lizards, and snakes. Modern-day birds are also *diapsids*, and scientists suspect that birds are the only direct descendants of dinosaurs.

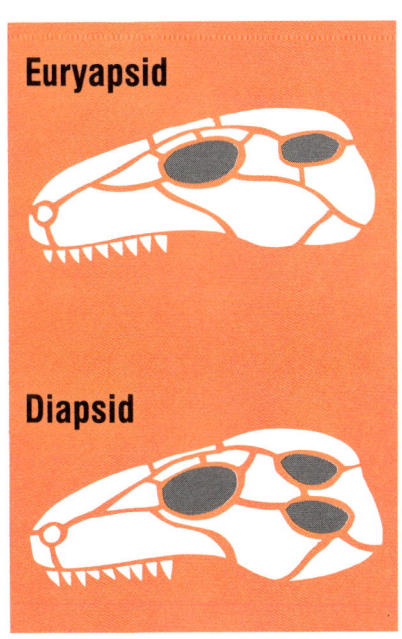

Euryapsid

Diapsid

It is also worth noting that the dinosaurs were very special *diapsids*. Not only did the dinosaur skull have the side openings we have already discussed, but it also had a large triangular opening above the eye sockets. Dinosaur skulls were the only *diapsid* skulls to have this extra opening. This difference sets dinosaurs apart. They were cousins to the other reptiles and close cousins to the other *diapsids*. But there the relationship ended.

The general scientific groupings you have just read about show the links that exist among members of the reptile family. But once we focus directly on our main interest—the land-living dinosaurs—the scientific names and groupings get more involved. The names aren't as hard as they sound; as long as you can keep in mind what they mean in English, you won't have any trouble.

To begin with, dinosaurs fell into two groups: the bird-hipped ones, or *ornithischians*, and the reptile-hipped ones, or *saurischians*. (*Ornith-* is Greek for "bird," and *saur-*is Greek for "reptile.") The bird-hipped dinosaurs were almost all herbivores, or plant-eaters, while the reptile-hipped group contained both meat eaters (carnivores) and plant-eaters.

The reptile-hipped group had two main subgroups: the *theropods*, or "wild-beast-footed ones," and the *sauropods*, or "reptile-footed ones." (*Pod* means "foot," by the way.) The deadly, toothy, famous *Tyrannosaurus* was a *theropod*, and the gigantic, slow-moving, well-known *Brontosaurus* was a *sauropod*.

The bird-hipped group had several subgroups: The *ornithopods*, or "bird-footed ones," usually had big, three-toed feet similar to those of the ostrich. The *ceratopsians*, or "horned and beaked ones," had a big, bony "frill" protecting their necks. They looked something like the rhinoceros. The best-known *ceratopsian* is *Triceratops*. The *pachycephalosaurs*, or "thick-headed reptiles," are only known to us from a very few fossil finds. Their skulls, as their name implies, are incredibly thick and bony. But there are hardly any definite remains of the rest of the *pachycephalosaur* skeleton, so not much is known about these dinosaurs. The *ankylosaurs*, or "fused reptiles," came in many shapes and sizes. As their

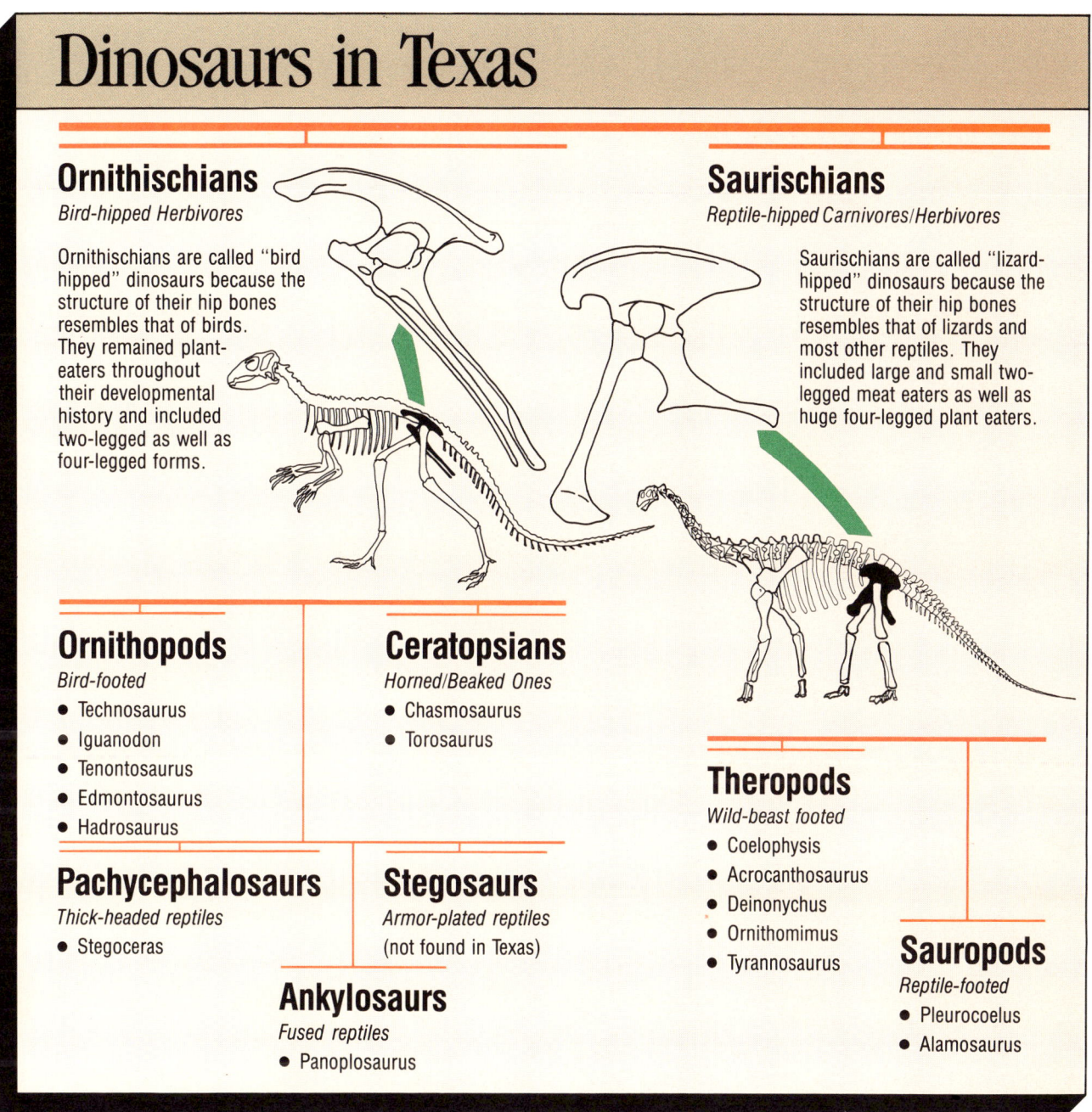

name indicates, they had all sorts of scary-looking horns and tusks all over their bodies. And last but not least, the *stegosaurs*, or "armor-plated reptiles," are well known to us due to the famous *Stegosaurus*, the dinosaur with the array of bony plates running the length of its backbone.

Types I-II

Type III

Type IV

DINOSAUR VALLEY STATE PARK at Glen Rose

Four kinds of distinctly outlined tracks are found in the park. Embedded in a lower Cretaceous limestone, part of an ancient seashore dated to 105 million years ago, they are clearly those of dinosaurs.

Types I-II These huge saucer-shaped and crescent like couplet (hind and front feet of the same animal) tracks were made by one of the giant, four-legged sauropods such as the Pleurocoelus, whose bones have been found in Wise and Blanco Counties.

Type III These slender-toed, sharp-clawed, tridactyl (three-toed) tracks were made by animals walking on their hind legs. This identifies the trackmaker as a theropod or carnosaur. The most likely candidate is Acrocanthosaurus, known from partial skeletons nearby.

Type IV These blunt toed, tridactyl tracks have been difficult to identify. A 1985 fossil discovery in West Texas of a large three-toed ornithopod is identified as an Iguanodon whose size and foot structure properly fit the Dinosaur Valley tracks.

Locations of Dinosaur Tracks

1. Denton County
2. Paluxy River, Somervell County
3. Lake Eanes, Comanche County
4. Cottonwood Creek, Hamilton County
5. S. San Gabriel River, Williamson County
6. Colorado River, Travis County
7. Garner Ranch, Kimble County
8. Blanco River, Blanco County
9. Comal County
10. Mayan Ranch, Bandera County
11. Hondo Creek, Bandera County
12. Davenport Ranch, Medina County
13. Hondo Creek, Medina County
14. Sabinal River, Uvalde County
15. Kinney County
16. Guadalupe River, Kerr County
17. Near Girvin, Pecos County

Part Two

Dinosaurs In Texas

Now we come to what you've been waiting for: the dinosaurs of Texas. Even though much of the state was covered with water at different times during the Mesozoic Era, dinosaur hunters have been able to find 16 different kinds, or species, of dinosaur here. Texas was quite a steamy, marshy place during the Mesozoic Era. The parts that weren't too deeply underwater provided the perfect hot, humid environment for the great thunder reptiles. Among them, the 16 species represent many of the known dinosaur families. They range from the small to the gigantic and from the quieter plant eaters (herbivores) to the most ferocious meat eaters (carnivores). And they have been found in three main locations: the two oldest species, 210 to 190 million years old, are from the Triassic Period and were found in the Panhandle; five species found in North Central and West Texas date from the older (Lower) Cretaceous Period and are 135 to 100 million years old; and nine species dating from the more recent (Upper) Cretaceous Period are 100 to 65 million years old and were found in the Big Bend area.

Let's start with the two oldest species, found in the Panhandle in rock formed during the Triassic Period:

Coelophysis

Coelophysis is one of the earliest dinosaurs known. Its bones have been found in Texas like the rest of the dinosaurs we will discuss. *Coelophysis* thrived all over Pangaea in its heyday. It was part of a family of slim, meat-eating, reptile-hipped dinosaurs called *coelurosaurs*. Its name comes from words meaning "hollow form," for its head and body were slender and it was very lightly built, weighing less than 100 pounds. But don't let that fool you; even though *Coelophysis* was only 8 feet long from head to tail (standing 3 feet at the hips), it was a meat-eating *theropod* ("beast foot"). This sleek little dinosaur stood on its two hind legs, using its tail for balance, and when it wanted to, it could run quite fast. Its narrow, lizard-like mouth had a menacing row of sharp little teeth. When it caught its prey—probably small, plant-eating dinosaurs as well as lizards and large insects—it could grip them with its three-toed hand claws, the better to bite down! In fact, some scientists believe this little dinosaur might even have been a cannibal, for some babies of the species at a site in New Mexico looked as if they had been eaten by the adults.

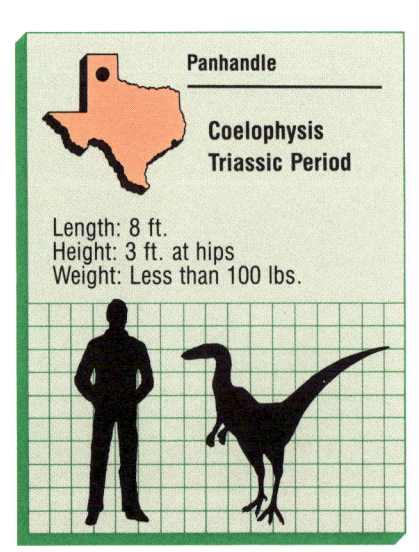

Panhandle

Coelophysis
Triassic Period

Length: 8 ft.
Height: 3 ft. at hips
Weight: Less than 100 lbs.

Technosaurus
(Represented here by Lesothosaurus)

Technosaurus, a bird-hipped plant eater only 4 feet long, probably provided many a meal for *Coelophysis*. It, too, was slight and light-boned, and stood on its hind legs. But there its resemblance to its meat-eating cousin ended. For one thing, it was an *ornithopod* ("bird foot"), so it didn't have vicious claws like *Coelophysis*. For another thing, its teeth were small and leaf-shaped, with gaps in-between—not exactly teeth that would stop a large attacker! Fragments of this tiny dinosaur's remains were found near Lubbock in the Panhandle, so it is named in honor of Texas Tech University. It was probably very similar in build and behavior to *Lesothosaurus*, an equally small herbivorous dinosaur found in Lesotho in southern Africa. *Lesothosaurus's* remains are in much greater supply than those of *Technosaurus*, however. So much of our knowledge about the Texas dinosaur comes from the African one. The two little beasts belonged to the same family of very early bird-hipped dinosaurs, the *fabrosaurs*.

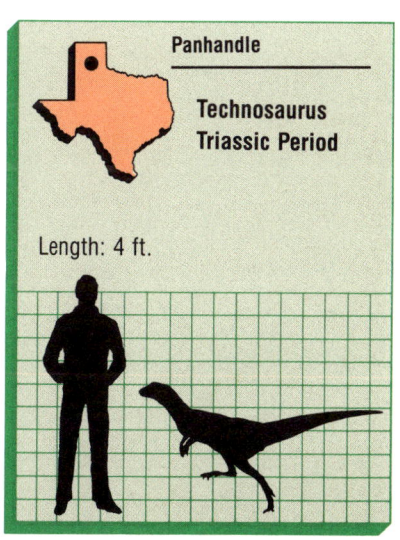

Panhandle

Technosaurus
Triassic Period

Length: 4 ft.

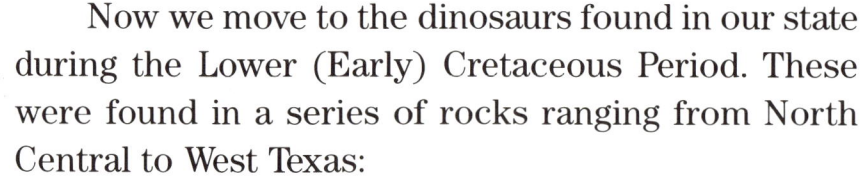

Now we move to the dinosaurs found in our state during the Lower (Early) Cretaceous Period. These were found in a series of rocks ranging from North Central to West Texas:

Acrocanthosaurus

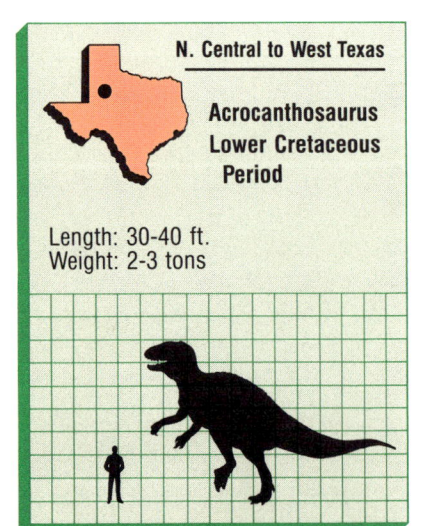

This strange animal's name means "high-spined reptile." That tells you right away how different it looked from the first two dinosaurs we discussed. It had a long, tall, skin-covered row of spines running down its back. Scientists think this ridge might have helped *Acrocanthosaurus* control its body temperature. When cold, for example, the dinosaur could turn so that its spiny ridge faced the sun. This would quickly warm up the blood circulating under the skin of the spiny ridge, and the beast would be ready for action. Not surprisingly, *Acrocanthosaurus* belonged to a family called *spinosaurids* (which is part of an even larger family called *carnosaurs*). But its spiny ridge was not the only thing to set it apart. It was very large—2 to 3 tons of dinosaur, 30 to 40 feet long! It was a *theropod,* so its claws were very fierce. And it had a large mouthful of jagged teeth. *Acrocanthosaurus* was quite a terrifying meat-eater.

Deinonychus

Speaking of terrifying meat-eaters, you can't find a much better example of one than this animal, whose name means "terrible claw." It was part of a dinosaur family called *dromaeosaurs*, which means "running reptile." It was perfectly built to be a fast runner, too: It had long, muscled legs and a straight tail that gave it good balance.

Deinonychus, a *theropod*, was small compared to some of its carnivorous cousins. It was 10 feet long from head to tail and was a two-footed (bipedal) animal standing 5 feet tall and weighing about 175 pounds. But it more than made up for small stature with its fierce, curved claws—especially those on its feet: Each foot had four toes. The first one was little, with a claw, or "toenail," that bent backwards and didn't rest on the ground. Instead it stuck out at an angle about where the animal's heel was. The third and fourth toes stretched forward, a little more the way they do on normal feet. The "toenails" on these two were a little larger than on the first. But the big "toenail" was the one in the middle—Number Two. It was 5

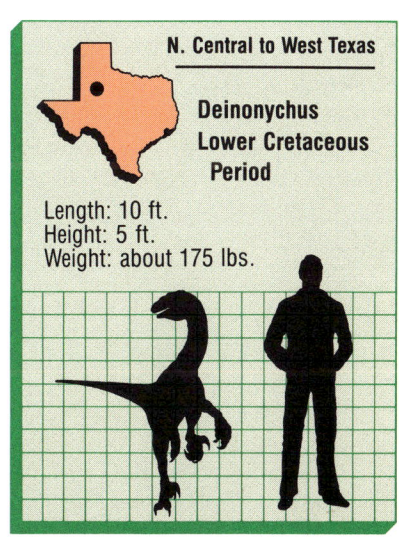

N. Central to West Texas
Deinonychus
Lower Cretaceous Period

Length: 10 ft.
Height: 5 ft.
Weight: about 175 lbs.

Fossil trackways provide information about dinosaur activities that skeletal remains do not. This illustration depicts tracks from Bandera County showing 23 sauropod trails, tightly clustered, with the animals moving in a single direction. Since the smaller tracks are located toward the middle of the cluster, some researchers think this herd was structured with the young guarded by adults on the perimeter. Scientists now know many kinds of dinosaurs traveled and grazed in herds or hunted in packs. No modern reptile exhibits such behavior.

inches long and sharply curved. It could move separately from the other toes—sort of the way a human thumb works.

To make a kill, *Deinonychus* would run very fast toward its prey with its hands outstretched. It would leap onto the victim, using its hand claws to grip and its tail to balance itself. Then it would shoot one of its feet out, fast as lightning, into soft, unprotected skin—often the stomach. The merciless "terrible claw"—the 5-inch one that is the beast's namesake—would then sink in. *Deinonychus* would grip, pull, and tear out the prey's innards. Most animals around the same size were no match for this fierce meat eater. And dinosaurs many times its size probably fell victim as well since scientists believe *Deinonychus* hunted in packs, like wolves. Several of them at once could have leaped onto the defenseless body of a large plant-eating dinosaur and made short work of it.

Pleurocoelus

This gigantic *sauropod* fits many people's definition of the classic dinosaur. It was 50 feet long and weighed an amazing 45 tons! It lumbered around on all four of its huge, rounded, lizard-like feet. It belonged to the *brachiosaurid* family. *Brachio-* means "arm"; this family of dinosaurs had arm bones that were extremely long compared to the size of the rest of the body. In addition, *Pleurocoelus*'s name means "side hollow." The beast had a very long neck and had hollows running along the sides of its backbone. It had a very small head at one end of this long neck and a massive body and tail at the other. It ate only plants. Because of its size, *Pleurocoelus* was fairly safe from attack, even from large *carnosaurs*. Nevertheless, some groups of fossil footprint trackways tell us that these great beasts and others like them found safety in numbers. They probably lived in herds, as many animals do today. That way they could encircle their young, protecting them from packs of predators.

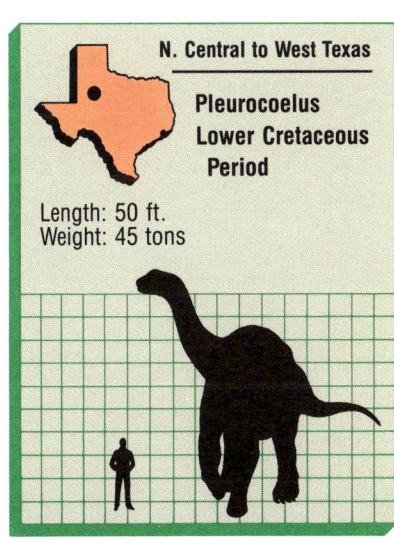

N. Central to West Texas
Pleurocoelus
Lower Cretaceous Period
Length: 50 ft.
Weight: 45 tons

Iguanodon

This *ornithopod* dinosaur was even larger (30 feet long and over 15 feet tall) and fatter (5 tons) than *Tenontosaurus*, but it lived earlier in geologic time. The two dinosaurs were very similar: They both moved around eating plants. *Iguanodon* and its relatives in the *iguanodontid* family also had bony "lips" and chewing teeth that were similar to *Tenontosaurus*— even though their family name means "iguana tooth," which sounds very different from "high-ridged tooth!" But when *Iguanodon* was found in Western Europe more than a hundred years ago, it was among the very first dinosaurs discovered. In those early days scientists were less precise in their naming than they are now.

Iguanodon was slower-moving than *Tenontosaurus*, but it had an impressive defense against enemies: a spiked thumb on each of its two hands. The animal spent most of its time wandering around on all-fours, but it could rise up and stab with its hands when necessary.

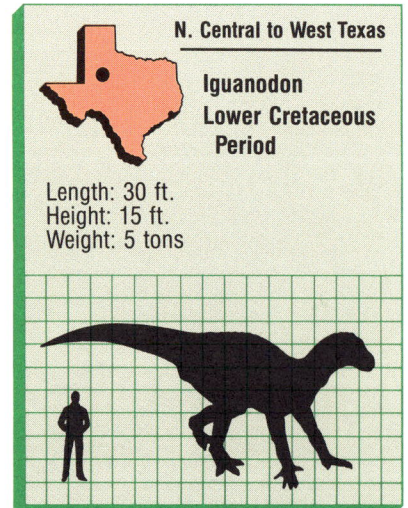

N. Central to West Texas
Iguanodon
Lower Cretaceous Period
Length: 30 ft.
Height: 15 ft.
Weight: 5 tons

Tenontosaurus

Tenontosaurus fossils have often been found in the same areas as those of *Deinonychus*. Scientists think it's likely that one of *Deinonychus*'s main victims was probably this fat, bird-hipped dinosaur. *Tenontosaurus* was an *ornithopod*, in many ways very much like *Technosaurus*, the tiny plant eater we discussed earlier. But *Tenontosaurus* weighed over a ton and was 15 feet long, over half of which was tail. It could run on its hind legs, but probably walked and browsed on all fours. It had a mouthful of heavy, crushing teeth and could grind down any and all vegetation, the tougher the better. Poor little *Technosaurus*, on the other hand, had gaps between its teeth and could do nothing more than swallow bite-size bits of plant matter.

Tenontosaurus means "sinew reptile." Sinews, or cords, in its tail enabled this beast to hold the tail out stiff and straight. The animal belongs to a family called *hypsilophodontids*, which is a very big word that means simply "high-ridged tooth." So you can see that one of "sinew reptile's" most important features was its wide, grinding teeth. They were hidden inside its mouth, behind bony, toothless "lips." The animal even had cheek pouches on each side of its face, where it could store partly chewed food.

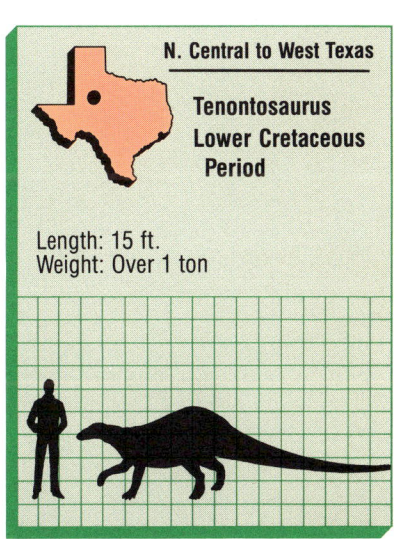

N. Central to West Texas

Tenontosaurus
Lower Cretaceous Period

Length: 15 ft.
Weight: Over 1 ton

By far the largest group of Texas dinosaurs comes from the Upper (Late) Cretaceous Period. Their remains have been found in rocks in the Big Bend area.

Alamosaurus

Big Bend Area

Alamosaurus Upper Cretaceous Period

Length: 70 ft.
Weight: 30 tons

This animal was another of the *Pleurocoelus* type—a huge, lumbering, four-legged, plant-eating *sauropod*. It was sleeker, though, since it had greater length (70 feet) and less weight (30 tons). *Alamosaurus* belonged to the *titanosaur* family, and since *titan* means "giant," it's clear the creature was known for its size. Unfortunately, though, scientists don't have much detailed knowledge about this dinosaur; not enough of its remains have been found.

Chasmosaurus

A large, bony plate sticks up from the back of this beast's head. Horns adorn its brow and nose. You probably have an idea already that *Chasmosaurus* is closely related to the well-known *Triceratops*. Both dinosaurs belong to the *ceratopsian* family. The word *ceratopsian* means "horned one"; *Chasmosaurus* certainly was horned! And besides that, the bony plate, or frill, that protected its neck made it look even stranger than *Triceratops*. This frill had a series of clefts, or openings, along its edge. It is those clefts that give *Chasmosaurus* its name (*chasmo-* means "cleft" or "ravine"). This *ceratopsian* was not huge, being only about 17 feet long. But it was thickly and powerfully built, weighing 2.5 tons. And even though it walked on all-fours, it stood several feet taller than we humans do.

Despite its scary-looking armor, however, *Chasmosaurus* and its kin were vegetarians that probably roamed the countryside in herds. Scientists think the armor might have served two important purposes: First, and most obviously, it protected the animals from the claws and teeth of attackers. Second, it was probably used by male *ceratopsians* as a form of display, to aid them as they marked out their territory. If a male crossed into a rival's territory, the two might run toward each other and butt heads, much the way certain kinds of sheep or deer do today. The winner of the head-butting contest would win control of the disputed territory and the females in that territory.

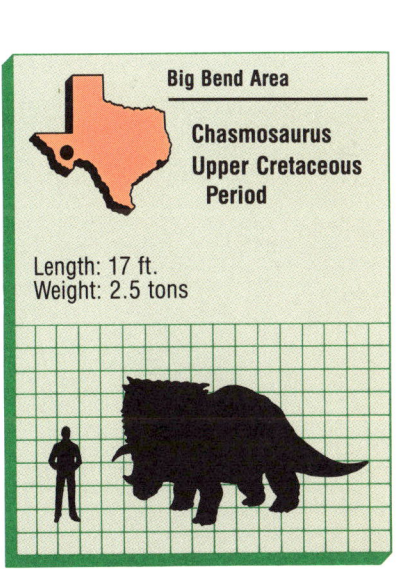

Big Bend Area

Chasmosaurus
Upper Cretaceous
Period

Length: 17 ft.
Weight: 2.5 tons

Edmontosaurus/ Hadrosaurus

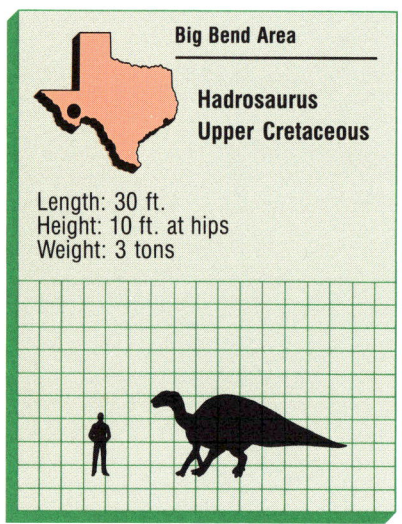

Big Bend Area

Hadrosaurus
Upper Cretaceous

Length: 30 ft.
Height: 10 ft. at hips
Weight: 3 tons

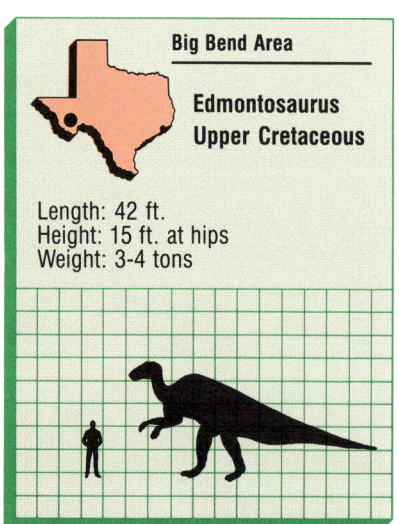

Big Bend Area

Edmontosaurus
Upper Cretaceous

Length: 42 ft.
Height: 15 ft. at hips
Weight: 3-4 tons

Here we have two *ornithopods* who are the better-known members of the *hadrosaurid* family. *Hadrosaurus*'s name means "big reptile," and *Edmontosaurus*'s name refers to Edmonton, Canada, near where it was first discovered. Both animals were two-legged, duck-billed vegetarians with massive rows of grinding teeth and cheek pouches to hold extra food. Their bodies were quite similar in build to that of the *Iguanodon*, except that neither *Hadrosaurus* nor *Edmontosaurus* had spiked thumbs to defend themselves with. The two beasts were similar in size, as well—*Edmontosaurus* being slightly larger (42 feet long, 15 feet at the hips, 3-4 tons) than *Hadrosaurus* (30 feet long, 10 feet at the hips, 3 tons). The *hadrosaurid* family was among the last of the bird-footed, herbivorous dinosaur groups to make an appearance on earth.

Ornithomimus

This dinosaur's name means "bird mimic." It and its fellow members of the *ornithomimosaur* family are also sometimes called "ostrich dinosaurs." From the following illustration, you can see how they came by their two names! They look very similar to the modern-day ostrich (which is probably descended from them). *Ornithomimus* was 12 to 15 feet long and stood 8 feet tall. It was not bird-footed, as you might expect from a dinosaur that looked like a bird. Instead it was a *theropod*, or "wild-beast foot." The last *theropod* that we talked about—*Deinonychus*— was very wild and very beastly indeed. But *Ornithomimus* was a *theropod* with no teeth and with no scary hooked claws. It did eat meat—but only the meat of insects and very small reptiles. It also ate some vegetation as well as any dinosaur eggs it could steal from unguarded nests. (Dinosaurs, like reptiles today, laid eggs—although there is evidence that a few species of dinosaur and some other ancient reptiles might have given live birth to their young.)

Like the ostrich, it had very long legs that gave it high speed, so it could make a fast getaway after raiding a nest, or race to catch a small lizard or winged insect. But unlike the ostrich, *Ornithomimus* didn't have feathers or wings. Where the ostrich uses its flightless wings to balance itself while running, *Ornithomimus* used its long, stiff tail. And *Ornithomimus* had an advantage over the ostrich: It could use its long, three-"fingered" hands to grasp a squirming lizard or a stolen egg.

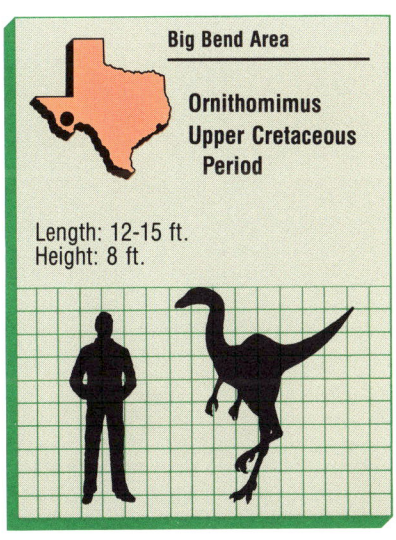

Big Bend Area

Ornithomimus
Upper Cretaceous
Period

Length: 12-15 ft.
Height: 8 ft.

Panoplosaurus

Big Bend Area

Panoplosaurus Upper Cretaceous Period

Length: 23 ft.
Weight: 3 tons

This animal's name means "fully armored reptile," so be prepared for a fairly strange appearance! It belonged to the *ankylosaur* family, and within that family it was in yet another group, the *nodosaurids*. *Ankylo-* means "fused," or "joined together," and *nodo-* means "nodular," or "bumpy." This fully armored, fused, bumpy dinosaur was not the most graceful creature ever to walk the Mesozoic earth! Despite its huge size (23 feet long, 3 tons), it waddled along on all-fours fairly close to the ground, more like a tank than an animal. It was bird-hipped and ate vegetation, and even though it had bony "lips," like some of the plant-eating dinosaurs we've already discussed, *Panoplosaurus* did not have good "grinders" in its mouth. Its teeth were small and leaf-shaped.

Stegoceras

Don't confuse this little dinosaur with the well known *Stegosaurus*, a huge animal with plates sticking up along its spine; the two couldn't be more different. *Stegoceras* means "horny roof." This 6-foot-long, 120-pound, bone-headed beast was a *pachycephalosaur*—which means "thick-headed reptile." On top of its head was 3 inches of solid bone. So it comes as no surprise that the male *Stegoceras* probably butted heads with its rivals (much as the *ceratopians* did).

Stegoceras's head was its most impressive feature. In other ways it was more typical: It walked on its hind legs, stood 22 inches at the hips, had serrated teeth for shredding plant matter, and had short arms. *Stegoceras* and its thick-headed relatives might well have lived very much as today's sheep and goats do in small groups in upland areas.

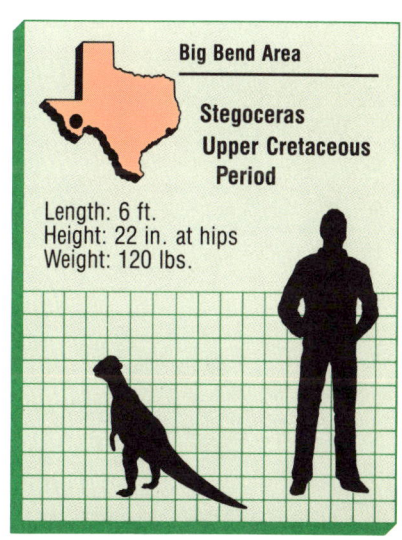

Big Bend Area

Stegoceras
Upper Cretaceous Period

Length: 6 ft.
Height: 22 in. at hips
Weight: 120 lbs.

Torosaurus

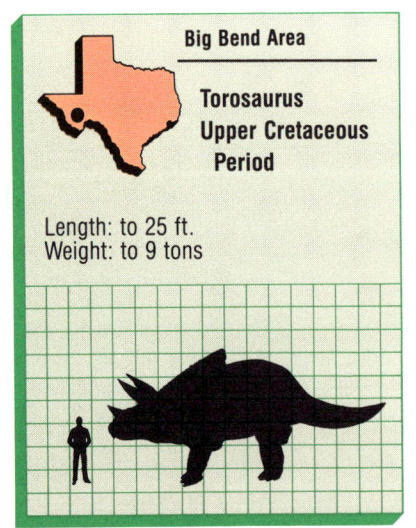

Big Bend Area
Torosaurus
Upper Cretaceous
Period

Length: to 25 ft.
Weight: to 9 tons

From the word *Toro-* ("bull"), you might have guessed that this *ceratopsian* looked something like a bull. *Torosaurus* was the largest of this family of horned, neck-plated dinosaurs. Like its cousin *Chasmosaurus*, *Torosaurus* walked on all-fours, ate vegetation, and traveled in a herd. But unlike the fairly small *Chasmosaurus*, this animal could weigh as much as 9 tons and could grow to be 25 feet long. The huge, curved, bony plate around the back of its head and the three horns on its face combined to make it look a lot like a bull. When it was charging a rival or an enemy, it must have looked like a bull with armor plating!

Some *Torosaurus* skull fossils are more than 8 feet long and 5 feet wide with brow horns 2 feet long.

Tyrannosaurus

Last but far from least, remains of the mighty *Tyrannosaurus* have been found in Texas. Its family, the *tyrannosaurids,* all lived up to the name of "tyrant reptile." *Tyrannosaurus* was the most ferocious and awesome meat eater of the dinosaur kingdom: a theropod weighing 6 to 7 tons, measuring more than 40 feet long from head to tail. The animal stood 18 feet high on thickly muscled legs and clawed feet. Its jaws were its most amazing feature: They were 3 feet deep and had 60 razor-edge teeth, some of them 7 inches long!

Although it could take big steps, *Tyrannosaurus* was way too large to be nimble and quick like its *theropod* cousin *Deinonychus.* Some scientists believe this huge *carnosaur* couldn't possibly have done all the running and chasing needed to catch all its prey. Instead, *Tyrannosaurus* might sometimes have eaten the flesh of already-dead dinosaurs.

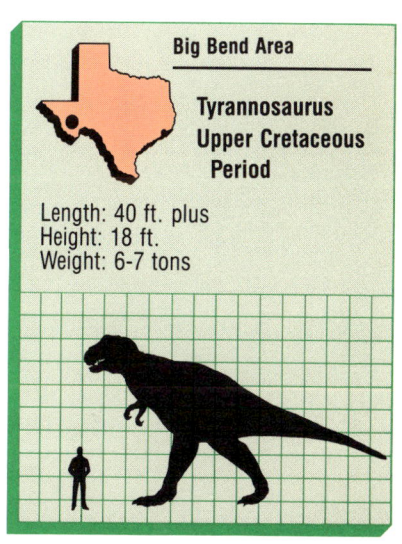

Big Bend Area

Tyrannosaurus
Upper Cretaceous
Period

Length: 40 ft. plus
Height: 18 ft.
Weight: 6-7 tons

Dinosaur Relatives Found in Texas

We have discussed Texas dinosaur fossil finds. We can't leave out the state's other reptile fossil finds: Some of the dinosaurs' closest and most impressive relatives lived here too.

Pteranodon/ Rhamphorhynchus/ Quetzalcoatlus

Comparisons give a startling picture of the Quetzalcoatlus size.

These animals with the long names all belonged to the *pterosaur* family of flying reptiles, close kin to the dinosaurs. They had membrane-covered wings and long, skinny jaws. They resembled modern-day bats more than birds, but even at that it wasn't a close resemblance. *Pteranodon*'s long beak was toothless and balanced by a huge crest at the back of the head. Logically enough, its name means "winged and toothless." It had a very short, stumpy tail and a fairly long neck. This flying reptile probably scanned the oceans in search of fish and had a mouth pouch in which to store them, much as the toothless pelican has today.

Rhamphorhynchus, on the other hand, had teeth in its beak, a fairly short neck, and a long tail that was strengthened by a series of bony rods. This animal, whose name means "narrow beak," was probably a fish eater too. But with all its teeth, it fished quite differently than *Pteranodon*, spearing or stabbing the prey rather than catching and storing it.

Quetzalcoatlus was given an Aztec name that means "feathered snake," even though the animal didn't have feathers. Its best-known feature was its size: It was by far the largest of the three *pterosaurs* discussed here with a wing span of 40 feet. It must have looked like an airplane when it flew!

Mosasaur skeleton on display at the Dallas Museum of Natural History. Discovered in 1979 on the shore of Lake Ray Hubbard, this 32 foot fossil once swam near Dallas when the area was part of the Gulf of Mexico.

Mosasaur

The *Mosasaur* was named for the Meuse region in Western Europe where it was first found. It was a marine, or water, reptile that often grew to lengths of 30 feet. It wasn't common only in Europe, for Texas was home to many *Mosasaurs* in the late Cretaceous Period. It had a mouthful of vicious teeth, a long tail that propelled it through the water, and webbed hands and feet that helped steer it. Scientists think *Mosasaur* was probably an ancestor to a very unusual and fierce-looking reptile that's alive today: the monitor lizard.

Phobosuchus

This was an unbelievably large Cretaceous crocodilian. So far, the only fossil remains ever found of *Phobosuchus* were unearthed in the Rio Grande region. Its name means "fearsome crocodile," and when you look at the following photograph, you'll see why. *Phobosuchus's* skull was 6 feet long; the rest of its body has not been found. But if it was built according to the same logic as today's crocodile, its head would probably have been followed by 40 feet of body! At that size, it could fairly well have fed on good-sized land dinosaurs such as *Hadrosaurus*.

Dr. Barnum Brown, R. T. Bird and Dr. Erich Schlaikjer examine a Phobosuchus skull.

This fossil, believed to be the largest Protostegid type turtle ever found, was discovered in the Eagle Ford shale formation near Gholson, Texas in 1971. It lived during the Upper Cretaceous period and swam the vast inland sea covering the Waco area about 75 million years ago. Outside measurements: flippers-13', length-12', shell-7'.

Protostega

Its name means "first roof," and it was a gigantic swimming turtle, ancestor to all the turtles still thriving today. It was 11 to 12 feet across with huge flippers. *Protostega's* shell was similar in quality to those of today's leatherback turtles.

Part Three

The State's Dinosaur Attractions

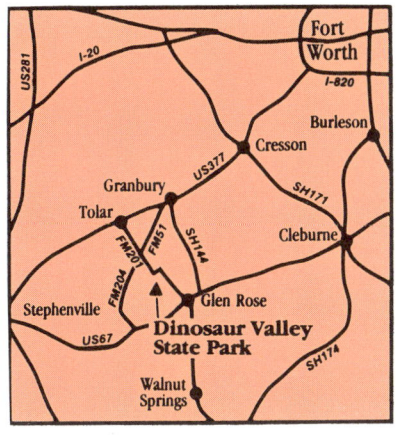

Dinosaur Valley State Park

Of course, it's great to read about all these creatures that used to inhabit our state (even though they have such long names!). But what about seeing some of their fossils? Texas gives you several good chances to do so, and one of the best is at Dinosaur Valley State Park at Glen Rose south of Fort Worth, which we described at the very beginning. There is something so very lively and vital about seeing animals' footprints. It's as though you just missed the two fighting dinosaurs as hunter chased hunted across the mud. You can almost imagine, looking at those tracks, that their makers will appear again.

The science of fossil study is called *paleontology*, and the paleontologist we have to thank for the preservation of the Glen Rose tracks and the creation of Dinosaur State Park is R.T. Bird. He did not find the first few tracks and trackways himself; local people did that. But when he came along, they were busy digging tracks up and selling them as souvenirs. They didn't realize that the real appeal of these fantastic footprints was in seeing them where they lay. Once dug up, the tracks could no longer reveal to paleontologists—or the public—many mysteries about the dinosaurs that made them. For instance, if scientists don't see footsteps in sequence, they can't tell how far apart the steps are. And if they can't tell that, they can't tell the size of the animal that made them. Also, if they don't see animals' tracks exactly as they were made,

Tommy Pendley taking an impromptu bath in the hindfoot track of dinosaur, Paluxy River site. Photo by R. T. Bird.

Allosaurus skeleton on display in dinosaur exhibit at the Panhandle-Plains Historical Museum, Canyon, Texas.

they can't tell what the animals were doing in relation to one another: whether one beast was walking alongside another, or walking faster than another, or attacking another. A lot would have been lost to us if R.T. Bird had not come along, realized the importance of the find, hired teams of men to help him unearth many hidden trackways, and then analyzed the trackways—so he could reveal to us new secrets of Texas's dinosaurs.

In addition to Dinosaur Valley State Park, there are museums of paleontology and natural history all over the state. You have probably visited some of them yourself. A few are mentioned here: The giant Texas *Mosasaurs* are a main feature at both the Dallas and Corpus Christi natural history museums. A marine cousin of the *Mosasaur*, called *Plesiosaur*, can be seen at the Witte Museum in San Antonio, along with the rhinoceros-like *Triceratops*, a cousin of the native Texans *Chasmosaurus* and *Torosaurus*. The Texas Memorial Museum in Austin displays not only a *Mosasaur* but also a *Dimetrodon*, a very early reptile belonging to a family called *pelycosaurs*. *Dimetrodon* had a huge, sail-like fin running down the middle of its back. This reptile and its kin died out just before the dinosaur era, but they shared some important qualities with early mammals and were probably their ancestors. The Fort Worth Museum of Science showcases remains of dinosaurs from out of state: the large and vicious *carnosaur Allosaurus*, and the plant-eating *Iguanodon's* close cousin *Camptosaurus*. At the Panhandle Plains Museum in Canyon resides another *Triceratops*, as well as another *Allosaurus*. The Texas Tech Museum in Lubbock has an *Allosaurus* replica as well. The Houston Museum of Science, also, features an out-of-state beast, a large, long-necked, plant-eating *sauropod* named *Diplodocus*. And the Strecker Museum at Baylor University in Waco displays the ancient sea turtle *Protostega*, as well as a fine *Plesiosaur*.

All of these fossils, and countless others, have been found, analyzed, and rebuilt by paleontologists like R.T. Bird. These are people trained to understand what the fossil remains are—and what their location in the earth reveals about them and their ancient lives.

Despite the knowledge and training of paleontologists, however, there are many questions they cannot answer for sure. A major question—one that has us all wondering—is the reason for the dinosaurs' disappearance at the end of the Mesozoic Era. Paleontologists have come up with many different possible reasons. Many have had long debates championing their theories.

One theory is that the dinosaurs' disappearance was caused by an immense natural disaster—a tremendous ball of matter from outer space hitting the earth, for example. If that had happened, it would have caused firestorms and thrown clouds of smoke and many tons of dust into the atmosphere. This would have cooled the earth by cutting off the sun's rays for quite a long period of time—years, probably. The cooling of the earth, and the clouding of the atmosphere, would first have caused the death of plants, which needed the sun's warmth in order to grow. The death of dinosaurs (and many other animals) would have followed, both because life-giving vegetation was dead and because dinosaurs needed the sun to warm their reptilian bodies and give them energy.

Another theory is that a more gradual series of events led to the dinosaurs' end. The oceans deepened, and many of the vast, shallow seas that had been part of the dinosaurs' habitat were gone for good. The earth gradually cooled, as well, causing death among these heat-loving creatures. The remaining beasts could no longer travel to warmer areas, since Laurasia and Gondwanaland (the northern and southern parts of the super-continent Pangaea) had split too far apart for that. And so, slowly, over time, due to these habitat changes and others like them, the dinosaurs vanished.

Both of these theories are plausible. It is also possible that a combination of natural disaster and gradual changes caused the mass disappearance. Perhaps a series of floods and earthquakes occurred and added to the effects of the general change in climate.

Many scientists do believe the end was caused by a number of factors, not just one or two. Since the world is so big and so complicated—and since dinosaurs had lived here so very long and so very successfully—it makes more sense to think that many factors together caused the great, powerful beasts to die out.

But whatever caused their destruction, our lives are richer for knowing that they existed. It is a real testament to human curiosity and intelligence, too, that we managed to make sense of the great puzzle of fossil bones—that we managed to fit those bones together and realize just how magical and amazing their owners must once have been.

In recent years geologists have discovered a thin layer of clay, rich in the chemical element iridium marking the boundary between Cretaceous rock and later deposits. Iridium is one of the rarest of metals on earth, but it is common in certain meteorites. Did a large meteorite strike the earth sending up a global cloud that changed the climate and caused extinctions? Scientists are working to find out.

R.T. Bird

Hunting for dinosaur fossils may sound like glamorous work. But before the excitement of finding something like a footprint or a bone comes a lot of difficult research and back-breaking labor—not to mention lots of luck! Being a dinosaur hunter is not for just anybody. You can tell that by studying Roland Thaxter Bird, the hunter who brought the Glen Rose dinosaur tracks to national attention. He was a pretty unusual character.

As a young man in the 1930s, R.T. Bird traveled across America on a motorcycle, investigating remote places and doing odd jobs to support himself. During his travels he happened to find a small fossil. When he went back home to New York, he showed it to a museum curator.

The curator was impressed with the young man's energy and with what he had found. He sent Bird out to work in Wyoming, at a site where fossils were being dug up. There and at another site in Colorado, the

Davenport Ranch site, Medina County 1940. Photo by R. T. Bird.

budding fossil hunter learned the scientific sleuthing methods he needed to know.

He used these methods to great effect in Glen Rose. But before he could do that, he had to find the Glen Rose site—and he found it completely by accident. Passing through Gallup, New Mexico, in the summer of 1938, he came upon some carved copies of three-toed dinosaur tracks on sale at a curio shop. He asked where they had been made, and the shop's owner told him about Glen Rose.

Bird got down to Glen Rose fast and began exploring the bed of the Paluxy River. Just as others had done earlier, he found the three-toed tracks of small carnivorous dinosaurs in several different spots. But his big find—one that no one else had yet made—was the track of another kind of dinosaur.

Around the three-toed tracks were what Bird at first thought were "potholes"—round holes about three feet across. He was so busy unearthing the small carnivores' tracks that he didn't see the importance of these "potholes." He even used them as containers for dirt he was digging up—without realizing that they, too, were dinosaur tracks.

They were the footprints of huge plant eaters. Until Bird discovered the tracks, scientists had thought the large sauropod dinosaurs that made them were strangers to land. But now he had provided concrete—or at least rock-solid—proof that these giants of the dinosaur kingdom didn't only live in water.

After many months' work with a large crew at the site, Bird had laid bare the series of tracks that has made Glen Rose really famous—tracks showing the carnivores' attack on their gigantic dinosaur relative.

R. T. Bird and carnivorous dinosaur tracks at the Paluxy River near Glen Rose.

Local Biology Teacher and Son Find Dinosaur Bones

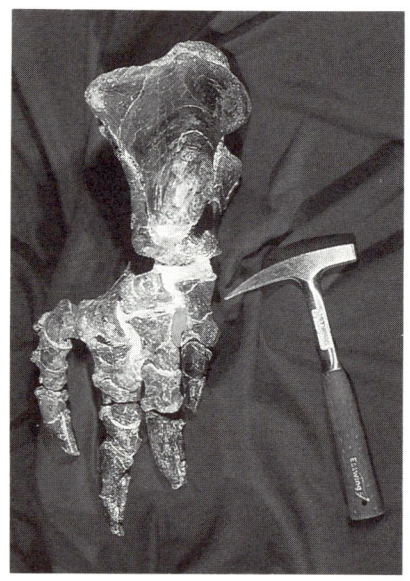

Illustration of Doss Ranch *Tenontosaurus*, a "bird-hipped" dinosaur of the Mesozoic Era.

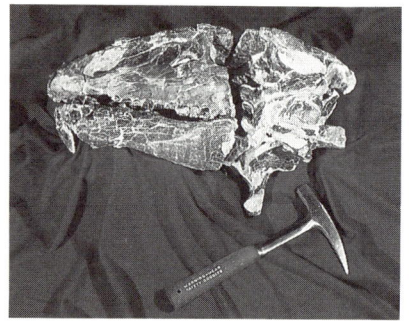

This *Tenontosaurus* hind foot was discovered in excellent condition in August 1988 at the Doss Ranch in Parker County.

The skull of a *Tenontosaurus*, a 110 million-year-old dinosaur, was found at the Doss Ranch in western parker County in August 1988.

Ted Williams and his seven-year-old son, Thad, often walk along stream beds looking for skulls and other specimens that Mr. Williams uses to teach biology to Millsap High School students, but what they found on August 5, 1988, will eventually be seen by many more people. Late that Friday morning, after turning a bend in the creek where water had cut deeply into the bank, they found some big bones while looking through brush.

Thad, a Millsap Elementary School second-grader, thought he saw teeth embedded in the side of an embankment. His father investigated further and recognized a skull shaped like that of a horse. Mr. Williams realized that the bones did not belong to deceased farm animals, but he was not certain that they were dinosaur bones. He and Thad scooped up a few of the bones and took them home. Upon bursting into their house young Thad exclaimed to his mother, "I've made a terrific discovery!"

On the following Monday morning, Mr. Williams and Mr. James H. Doss, the landowner, decided to call the Fort Worth Museum of Science and History. Museum chief associate science curator Jim Diffily visited the site on Tuesday and identified the fossils as Cretaceous reptile bones but could not give further positive identification. On Wednesday, Mr. Diffily took several of the bones, including a portion of the skull, to colleagues at the Dallas Museum of Natural History where Charles Finsley identified them as similar to a partial *Tenontosaurus* skeleton in the Dallas

Museum's collection. Later, paleontologists from Southern Methodist University and Tarleton State University, where Mr. Williams received his undergraduate and Master of Biology degrees, aided in the precise determination of the find as that of *Tenontosaurus*.

Both of the Williamses are delighted that their find will be seen by millions of Museum visitors. "It's exciting to discover something as rare as this, and even more satisfying to know how significant this find is," said Mr. Williams.

—*Fort Worth Museum of Science and History*

Fort Worth Museum of Science and History staff members carefully look for Doss Ranch *Tenontosaurus* remains.

Afterword

As you read in the foreword of this book, there are fossil discoveries out there just waiting to be made. And many of those who make such discoveries are not paleontologists but curious people, like you. So keep your eyes open. Finding a fossil is not as unlikely as it sounds.

In 1985, a student at Tarleton State University was exploring an area near Proctor Lake. There he found the fossilized remains of a *hypsilophodont*, a dinosaur related to the *Tenontosaurus* described earlier. And in 1988, a woman walking her dog found the fossil of *Xiphactinus*, an 80-million-year-old fish that was 10 feet long. Imagine how thrilled these two people must have been when they made their finds. They weren't "experts"—just observant and interested folks.

Our world is constantly changing, and for those who take notice, there are always things to discover.

Glossary

Acrocanthosaurus (a kro kan tho SAW ruhs)—name means "very spiny lizard." The Acrocanthosaurus was a theropod with a relatively small head and an overall length of about 40 feet.

Alamosaurus (a luh mo SAW ruhs)—name means "Alamo lizard." The Alamosaurus was one of the last sauropods to live in North America. It was close to 70 feet long and weighed 30 tons.

Allosaurus (AL o sawr us)—a theropod, name means "other lizard." Weight: 1-2 tons. Length: possibly 42 feet.

Amphibian (am FIB e on)—cold-blooded, chiefly egg-laying vertebrate who is adapted for life both on land and in water. Frogs and salamanders are amphibians.

Anapsids—group of the most primitive reptiles, to which the turtles and tortoises belong. They have no opening in the skull other than the eye sockets.

Ankylosaur (an KILE o sawr)—name means "fused lizard." Ankylosaurs were the armored dinosaurs. They had low, squat, heavy, barrel-shaped bodies. They were built like living military tanks with a flexible armor of bony slabs, plates, and spikes set in tough skin. The largest weighed about as much as a small elephant. They were another of the ornithischians.

Archeology (AR ke OL o je)—the scientific study of history from the remains of past human activities, such as burials, buildings, tools, and pottery.

Archosaurs (AR ko sawrs)—name means "ruling reptiles." Archosaurs were the reptile group to which the dinosaurs, pterosaurs, and crocodiles belonged. They ruled during most of the Mesozoic Era.

Bedrock—solid unweathered rock lying underneath the Earth's layer of loose soil, earth or rock.

Brachiosaurid (BRAK ee o sawr id)—name means "arm lizard." The brachiosaurids were sauropods who were built like immense giraffes. Some of them may have weighed as much as 20 big elephants! Brachiosaurids probably lived on every continent except Antarctica.

Brontosaurus (BRON toe sawr us)—a sauropod "thunder reptile." Length: 70 feet. Shoulder height: 14.5. 33 tons.

Camptosaurus—an ornithopod, belonged to a group of dinosaurs that first flourished during the Jurassic Period. These dinosaurs had bird-like pelvises and horny beaks used for eating plants. Camptosaurus ("bent lizard") was a small dinosaur, being only 5 to 10 feet in length. It had hoof-like feet and hands, which were useful for pulling and holding foliage but were useless for defense.

Carnosaur (kar nuh SAWR)—name means "flesh lizards." The carnosaurs were theropods, meat-eating saurischian dinosaurs who walked on two legs. They were usually heavily built and had short necks and arms but powerful legs and very large heads.

Ceratopsian (sair a TOP see yan)—the ceratopsians were ornithischian (bird-hipped) dinosaurs. They were one of the last types of dinosaurs to appear and were very abundant. Most were four-legged; they had huge heads and their protection was their very formidable horns. The best known is the Triceratops.

Chasmosaurus—(kaz mo SAW ruhs) "cleft or ravine reptile"; a ceratopsian, Chasmosaurus was 17 feet long and thus smaller than its relative, the Triceratops.

Coelophysis—(see LAW fih sihs) one of the earliest and most primitive dinosaurs. Coelophysis was a Triassic coelurosaur with an elongated head filled with fine, sharp teeth for tearing flesh. Coelophysis ("hollow form") was 8 feet long and 3 feet high but weighed only 100 pounds, partly as a result of having light-weight, hollow bones.

Coelurosaurs—("hollow-tailed reptile") meat-eating bipedal saurischian dinosaur, generally lightly built, with long neck and arms, slender legs, and a relatively small head.

Comanchean Series (ka MAN CHEAN)—the Cretaceous has two divisions in Texas: the Lower Cretaceous (early) and the Upper Cretaceous (late). The Lower Cretaceous is also known as the Comanchean Series.

Continental Sediment Deposit—material that has been deposited or laid down on land by settling out from the medium which was transporting it, such as water or air.

Cretaceous (kri TA shas)—the third and youngest division of the Mesozoic Era, lasting from about 135 million to 65 million years ago.

Deinonychus (dai NAW nee kuhs)—name means "terrible claw." A coelurosaur (a carnivore) who measured 10 feet in length and must have been a swift runner. Deinonychus had a unique weapon: the second toe of each foot had a huge, sickle-shaped claw that remained off the ground. When he kicked his leg backwards, this claw could rake across a foe with tremendous force.

Diapsids—a reptile group, including the dinosaurs, as well as crocodiles, lizards, snakes, and descendants of the birds, characterized by having a pair of openings immediately behind the eye sockets.

Dimetrodon (dye MET roh don)—a carnivorous, primitive lizard that lived during the late Permian Period. It possessed a sail, which stretched the length of its back and helped to regulate body temperature. When too cold, it would turn broadside to the sun to absorb heat. Dimetrodon was about 11 feet long and lived in what is now Texas.

Dinosaur (DI no sawr)—"extinct, prehistoric reptile" has long been the accepted, if incomplete, definition of the dinosaur. However, during the last ten years or so, it has been argued that the dinosaur should be in a class by itself, a class equal to that of reptiles. Some scientists have even put forth the theory that maybe dinosaurs were warm-blooded like mammals. Perhaps we will never know all the secrets of the wonderful animals who lived so long ago.

Diplodocus (DIP lo DOH kus)—name means "double beam." The Diplodocus was a sauropod. It was one of the longest dinosaurs with a total length of 85 feet or more, much of which was snaky neck and whiplash tail. Its weight was only 10-12 tons. It had peg-like teeth that clustered closely together to form a fringe around the front of the mouth only.

Dromaeosaurs—name means "running reptile." The huge, sickle-like claw on the second toe of the foot is the most characteristic feature of the dromaeosuar family of dinosaurs.

Duckbill—The duckbills were ornithopods who got their name from their broad, toothless beaks. Yet they had more teeth than any other dinosaur; their back jaws were literally crammed with them. Some of them stood as tall as a house; others were quite tiny. They were among the most abundant dinosaurs in North America during the Cretaceous period.

Edmontosaurus (ehd mahn to SAW ruhs)—a large animal, up to 42 feet long. Edmontosaurus had about a thousand strong teeth in the cheek region. The low area on top of the skull near the front might have been covered with loose skin which could have been inflated to make a loud bellowing call.

Era—a division of geologic time. An era includes one or more periods.

Euryapsids—a reptile group, exclusively aquatic and now extinct (e.g. plesiosaurs and ichthyosaurs) characterized by a single opening high up on the side of the skull behind the eye sockets.

Fabrosaurs—early bird-hipped dinosaurs.

Formation—A rock unit used by geologists; layers of sedimentary rock deposited under fairly uniform conditions during a limited interval of time. A formation is established and identified on the basis of its definite physical and chemical characteristics. Usually, formations are given names that reflect the type of rock (such as clay, shales, or limestone, etc.) and their location. Examples of formations in Texas are Goliad Clay, Beaumont Clay, Carrizo Sand, and Glen Rose Limestone.

Fossil—the remains or traces of plants and animals that lived long ago and have been buried by natural causes and preserved in the Earth's crust. Fossils include tracks and imprints of the long dead animals.

Frill—a long, back-swept bony crest, jutting out above the neck and shoulders of the ceratopsian dinosaurs.

Geology (je OL a je)—study of the rocks of the Earth's crust in order to learn more about the origin, structure and history of our world and its past inhabitants.

Geological Time Scale—the subdivisions of geologic time shown on a chart or table with the first or oldest era of time at the bottom and the last or youngest at the top.

Gondwanaland (gond WA na land)—southern super continent of early Mesozoic times, composed of what was later South America, Africa, India, Antartica, and Australia.

Gulf Series—the Cretaceous has two divisions in Texas: the Lower Cretaceous (early) and the Upper Cretaceous (late). The Upper Cretaceous is also known as the Gulf Series. (See Comanchean Series and Cretaceous listed above.)

Hadrosaurids—family of ornithopods appearing in the middle Cretaceous period.

Hadrosaurus (ha dro SAW ruhs)—moderately large ornithopod of late Cretaceous times. Also called duckbill.

Hypsilophodontids (hip si LOFE o don tids)—small, slim, and speedy ornithopods. Their kind were among the fastest runners of the Age of Dinosaurs.

Ichnite (IK nite)—fossil footprint.

Iguanodon (ih GWAHN o dahn)—name means "iguana tooth." Iguanodons were ornithopods. Length: up to 30 feet. Weight: up to 5 tons.

Iguanodontids—ornithopod dinosaur family of Lower Cretaceous period.

Jurassic (joo RAS is)—the middle period of the Mesozoic Era lasting from about 180 million to 135 million years ago.

Laurasia (Lo RA zha)—northern super continent of early Mesozoic times, composed of what would eventually be North America, Europe, and Asia.

Lesothosaurus (le SOO toh SAW rus)—ornithischian dinosaur; small ornithopod (three feet long), resembling Coelophysis in its general appearance. Lesothosaurus seems to have been one of the most primitive of all ornithischian dinosaurs because it appears not to have had cheeks.

Mammal (MAM al)—vertebrates who use lungs to breathe, have warm blood and hairy skin. The females give birth to their young alive and feed their young with milk from mammary glands.

Mammoth (MAM oth)—a type of elephant, now extinct. It had extra long tusks and very thick hair.

Marine life—native to the sea; organisms that live in the sea.

Mesozoic Era (MEZ o ZO ik)—name means "middle life." The period of time in Earth's history between 225 million and 65 million years ago. It was the era when the reptiles ruled the land. The Mesozoic was divided into the Triassic, Jurassic, and Cretaceous periods.

Mosasaur (mo zuh SAWR)—extinct marine lizard. The mosasaurs were large, fish-eating relatives of modern day lizards. They flourished in the great Cretaceous seas that covered much of Texas during that time.

Nodosaurid—a subdivision of the ankylosaur family.

Nodosaurus (NO doe sawr us)—name means "node lizard." The nodosaurus was one of the ankylosaurs (armored dinosaurs). It had alternating rows of large and small plates down the back and flanks. Length: up to 18 feet.

Ornithomimus (awr nih tho MAI muhs)—name means "bird mimic." The ornithomimus was one of the "ostrich dinosaurs," (ornithomimosaur family) the tall, sharp-eyed, toothless theropods. It grew to be over eleven feet long, more than half of which was tail.

Ornithopod (or ni THOP od)—name means "bird feet." Ornithopods made up one of the major groups of ornithischian (bird-hipped) dinosaurs and the only group capable of walking or running on their hind legs. The earliest ornithopods were rather small, no heavier than man, but the later, heavier ones might weigh as much as an elephant.

Ornithischia (or nih THISS kee ah)—one of the two great orders of dinosaurs. Ornithischians were characterized by hip bones arranged like those of birds. Many of them had horny, toothless beaks, powerful grinding teeth and pouches in their cheeks. Many of them also had a lattice of bony tendons that reinforced the spine.

Outcrop—geological term: a place where the bedrock is exposed at the surface.

Pachycephalosaurs (PAKKi SEFF aloe SAW rus)—name means "thick headed reptiles." Another group of Cretaceous ornithischians. With their skulls encased in a thick covering of bone, these dinosaurs may have butted their heads together during mating season rivalries.

Paleontology (PA le on TOL e je)—the science that deals with the study of ancient forms of life or of fossil organisms.

Pangaea (Pan JEE ah)—the one original supercontinent, presumed to have existed 200 million years ago.

Panoplosaurus (pa naw plo SAW ruhs)—name means "fully armored lizard." It was one of the last North American nodosaurids. It had hard plates encased in its back and bony armor fused to its skull. The Panoplosaurus was up to 23 feet long and weighed between 3 and 4 tons. It lived from Alberta in Canada to our Texas.

Pelycosaurs—the earliest mammal-like reptile called ('sail reptiles') that were first found in rocks of the late Carboniferous age. Primitive member of the synapsid group of reptiles.

Phobosuchus (fo bo SOOK us)—name means "fear crocodile." A large crocodile with its length up to 50 feet. These great beasts were abundant in the Cretaceous period in what was later Texas.

Plesiosaurs—('ribbon reptiles'); large-bodied, aquatic reptile of the Mesozoic Era, which swam using large paddle-like legs.

Pleurocoelus (plu ro SEE luhs)—plant-eating sauropod from the Lower Cretaceous period of Texas.

Prehistoric (PRE his TOR ik)—of or belonging to the period before written history.

Protostega—name means "first roof." A gigantic swimming turtle eleven to twelve feet across with huge flippers.

Pteranodon (tair AN o don)—name means "wing without tooth." The pteranodon was a pterosaur with a body about the size of a large turkey; it had a long toothless beak.

Pterosaur (tair o SAWR)—name means "winged lizard." Pterosaurs were flying reptiles of Mesozoic times. One might call them the "air force" of the archosaurs.

Quetzalcoatlus (ket sol ko AT lus)—name means "legendary or mythical bird." The quetzalcoatlus was a giant pterosaur.

Reptile (REP til)—Reptiles are cold-blooded, air-breathing scaly-skinned animals who generally lay eggs with shells. Snakes, lizards, crocodiles, and turtles are all reptiles.

Rhamphorhynchus (RAM foe RINK us)—name means "Prow Beak." The rhamphorhynchus was a small, long-tailed pterosaur.

Saurischian (sawr ISS kee yan)—dinosaur characterized by the arrangement of its hip bones. It was called "lizard-hipped" because of the resemblance of its pelvic bone structure to that of lizards.

Sauropoda (sawr a PODE a)—name means "lizard foot." The sauropods were enormous, plant-eating, four-legged saurischian dinosaurs. See Brontosaurus. Alamosaurus and Pleurocoelus were found in Texas. The Pleurocoelus tracks are outstanding at Glen Rose.

Sediment (SED e ment)—material that has been deposited by settling out of some moving medium, such as air or water.

Sedimentary Rock—rocks formed by the accumulation of sediments over a long period of time.

Serpentine (SUR pen teen)—snakelike.

Spinosaurus—a 39 foot-long theropod who lived in late Cretaceous Egypt. Much like the Permian lizard, Dimetrodon, he had a sail which ran the length of his back, probably acting as a sort of radiator.

Spinosaurids—family of high spined carnosaurs. Examples: Acrocanthosaurus and Spinosaurus.

Stegoceras (steh GAH seh ruhs)—a medium-sized pachycephalasaur, Stegoceras was possibly 6.5 feet long. The high dome on the skull was not so large in juveniles, but it became relatively very thick only in older animals.

Stegosaur (STEG o sawr)—"armor-plated reptile." One of the categories of ornithischians, the best known is the Stegosaurus.

Stegosaurus (STEG o sawr us)—name means "covered lizard" or "roof lizard." The Stegosaurus is the biggest and perhaps best known plated dinosaur. It grew to be 30 feet long and weighed up to 2 tons.

Strata (STRA ta)—plural of stratum. A stratum is a single bed or layer of sedimentary rock. (See sedimentary rock above.)

Synapsids—extinct group of reptiles; ancestral to the mammals and characterized by a single (lower) opening in the side of the skull behind the eye sockets.

Technosaurus (tehk no SAW ruhs)—a bird-hipped plant eater only four feet long that was an ornithopod.

Tenontosaurus (teh nahn to SAW ruhs)—name means "sinew lizard." The Tenontosaurus was a ornithopod with a very long tail, much longer than the rest of its body. It also had long forelimbs and probably walked on all fours much of the time. It weighed around 1 ton and was a little over 15 feet long.

Terrestrial (te RES tre el)—living on, or growing in the earth or land.

Tetrapod (TET ra pod)—very simply put, an animal with four feet.

Texas High Plains—Part of the Great High Plains which lie to the east of the base of the Rocky Mountains and extend into Northwest Texas. Also known as the Staked Plains or the Llano Estacado.

Theropoda (thair o PODE a)—name means "beast feet." Theropoda (theropod) was one of the two great orders of saurischian or "lizard-hipped" dinosaurs. The theropods were two-legged flesh-eaters.

Titanosaur—a poorly defined, mainly Cretaceous group of sauropods: tail vertebrate have cupped front surface, possibly armoured.

Torosaurus (toh ro SAW ruhs)—name means "bull lizard" or "piercing lizard." The Torosaurus was a horned dinosaur, the largest of the long-frilled ceratopsians. It was 25 feet long and weighed 8-9 tons. It ranged from Montana to Texas.

Trans-Pecos—the triangular "panhandle" of Texas west of the Pecos River, bounded on the north by New Mexico and on the south by the Republic of Mexico.

Triassic (tri AS ik)—the earliest period in the Mesozoic Era, lasting from about 225 to 180 million years ago.

Triceratops (try KERR a tops)—name means "three-horned face." The Triceratops was one of the last and largest ceratopsians, weighing in at close to 6 tons and measuring up to 30 feet in length. Nearly a third of its length was a massive head with a short frill and formidable brow horns.

Tyrannosaurus (tih ran no SAW ruhs)—name means "tyrant reptile." The Tyrannosaurus was among the biggest and most powerfully armed theropods. It measured nearly 40 feet in length, towered between 15 and 20 feet tall standing on its hind legs and weighed 7 tons. Its saw-edge teeth were over 7 inches! (Also see carnosaur above.)

Bibliography

Andrews, Roy C. *In the Days of the Dinosaurs*. New York: Random House, 1959.

Arbingast, Stanley A., Lorrin G. Kennamer, Robert H. Ryan, James R. Buchanan, William L. Hezlep, L. Tuffly Ellis, Terry G. Jordan, Charles T. Granger, and Charles P. Zlatkovich. *Atlas of Texas*. Austin: Bureau of Business Research, 1976.

Bird, Roland T. *Bones for Barnum Brown*. Fort Worth: Texas Christian University Press, 1985.

Elting, Mary. *The Macmillan Book of Dinosaurs and Other Prehistoric Creatures*. New York: Macmillan Publishing Co., 1984.

Glut, Donald F. *The New Dinosaur Dictionary*. Secaucus: Citadel Press, 1982.

Lambert, David. *A Field Guide to Dinosaurs*. New York: Avon Books, 1983.

_____. *The Field Guide to Prehistoric Life*. New York: Facts on File Inc., 1985.

Langston, Wann, Jr. "Nonmammalian Comanchean Tetrapods." Volume 8 in *Geoscience and Man*, edited by Bob L. Perkins. Baton Rouge: Louisiana State University, 1974.

Matthews, William Henry, III. *Texas Fossils: An Amateur Collector's Handbook*. Austin: Bureau of Economic Geology, 1971.

Mossman, David J., and William A. S. Sarjeant. " The Footprints of Extinct Animals." *Scientific American* 248 (1) (January 1983): 74-85.

Norman, David. *The Illustrated Encyclopedia of Dinosaurs*. New York: Crescent Books, 1985.

_____. *When Dinosaurs Ruled the Earth*. New York: Exeter Books, 1986.

Rao, Anthony. *The Dinosaur Coloring Book*. New York: Dover Publications Inc., 1980.

Sattler, Helen Roney. *The Illustrated Dinosaur Dictionary*. New York: Lee and Shephard Books, 1984.

Wilford, John Noble. *The Riddle of the Dinosaur*. New York: Alfred A. Knopf, 1985.

Williams, John E., "Images of Dinosaurs." *Texas Parks and Wildlife*, Volume 45 No. 8 (August 1987), Austin, Texas.

Zumberge, James H., and Clemens A. Nelson. *Elements of Physical Geology*. New York: John Wiley & Sons Inc., 1976.

Index

A

Acrocanthosaurus, characteristics of, 30
age of fossils, radiometric dating of, 15
age of the earth, theories about, 15-16
Alamosaurus, characteristics of, 36
Allosaurus, 51
    skeleton of, 50
anapsids (group of reptiles), 22
ankylosaurs, 40
    characteristics of, 24, 25
Austin, Texas, 51

B

Bandera County, fossils in, 32
Big Bend area dinosaurs, 27
    Alamosaurus, 36
    Chasmosaurus, 37
    Edmontosaurus, 38
    Hadrosaurus, 38
    Ornithomimus, 39
    Panoplosaurus, 40
    Stegoceras, 41
    Torosaurus, 42
    Tyrannosaurus, 43
Bird, R.T., 49, 50, 51
    career of, 54
    work at the Glen Rose site, 55
bird-hipped dinosaurs. *See ornithischians*.
brachiosaurids, 33

C

Camptosaurus, 51
Canyon, Texas, 50, 51
carnosaurs, 30, 43, 51
Cenozoic Era, 19
ceratopsians, 37, 42
    characteristics of, 24, 25
Chasmosaurus, 51
    characteristics of, 37
Coelophysis, characteristics of, 28
coelurosaurs, 28
continents
    breaking up of, 18, 20-22
    Gondwanaland, 20
    Laurasia, 20
    Pangaea, 18
Corpus Christi, Texas, 50
Cretaceous period, 19, 20

D

Dallas, Texas, 50
dating of fossils, radiometric, 15
Davenport Ranch, fossils at, 54
Deinonychus, characteristics of, 31-32
diapsids (group of reptiles), 23
Dimetrodon, characteristics of, 51
dinosaur, origin of the word, 15
Dinosaur Valley State Park, 49
    tracks found in, 26
Diplodocus, 51
disappearance of dinosaurs, 22
    theories of, 51-54
dromaeosaurs, 31-32

E

earth, geologic changes of, 18
"Earth Clock" (illustration), 16-18
Edmontosaurus, characteristics of, 38
euryapsids (group of reptiles), 23
exhibits of dinosaurs in museums, 50-51
extinction of dinosaurs, 22
    theories of, 51-53

F

fabrosaurs, 29
Fort Worth Museum of Science, 51
fossils
    at Bandera County, 32
    at Davenport Ranch, 54
    definition of, 13
    formation of, 13
    at Gholson, Texas, 48
    at Glen Rose, Texas, 11-13, 15, 55
    important finds in Texas (map), 14
    preservation of, 49-50
    at Proctor Lake, 56
    radiometric dating of, 15

G

geologic changes of the earth, 18
geologic time scale
    "Earth Clock" for showing, 16-18
    eras when dinosaurs lived, 18-20
Gholson, Texas, fossils at, 48
Glen Rose, Texas
    dinosaur park located near, 49
    picture of dinosaur tracks, 12
    reconstruction of events at (illustration), 13
    site of dinosaur tracks and fossils, 11, 13, 55
Gondwanaland (continent), 20

H

hadrosaurids, 38
Hadrosaurus, characteristics of, 38
Houston Museum of Science, 51
humans, time of existence on earth, 16-18
hypsilophodontids, 35

I

iguanodontids, 34
Iguanodon, 51
    characteristics of, 34
iridium, as a cause of extinction of dinosaurs, 53

J

Jurassic period, 19, 20

L

Laurasia (continent), 20
Lesothosaurus, characteristics of, 29
locations of dinosaurs in Texas. *See* Big Bend area dinosaurs; Glen Rose, Texas; North Central and West Texas dinosaurs; Panhandle area dinosaurs.
Lubbock, Texas, 51
    dinosaur found near, 29

M

Medina County, fossils in, 54
Mesozoic Era, 19
    breaking up of continents during, 20-22
    important fossil finds in Texas (map), 14
    three parts of, 20
Mosasaur
    characteristics of, 46
    in museum exhibits, 50, 51
museum exhibits of dinosaurs, 50-51

N

nodosaurids, 40
North Central and West Texas dinosaurs, 27
    Acrocanthosaurus, 30
    Deinonychus, 31-32
    Iguanodon, 34
    Pleurocoelus, 33
    Tenontosaurus, 35

O

ornithischians
    characteristics of, 24, 25
    Edmontosaurus, 38
    Hadrosaurus, 38
    Iguanodon, 34
    Panoplosaurus, 40
    Technosaurus, 29
    Tenontosaurus, 35
ornithomimosaurs, 39
Ornithomimus, characteristics of, 39
ornithopods
    characteristics of, 24, 25
    Edmontosaurus, 38
    Hadrosaurus, 38
    Iguanodon, 34
    Technosaurus, 29
    Tenontosaurus, 35

P

pachycephalosaurs, 41
    characteristics of, 24, 25
paleontology, definition of, 49
Paleozoic Era, 19
Paluxy River, dinosaur tracks at, 12, 55
Pangaea (continent), breaking up of, 18, 20-22
Panhandle area dinosaurs, 27
    Coelophysis, 28
    Technosaurus, 29
Panhandle Plains Historical Museum, Canyon, Texas, 50, 51
Panoplosaurus, characteristics of, 40
pelycosaurs, 51
Pendley, Tommy (photograph), 49
Phobosuchus, 47
Plesiosaurs, 51
Pleurocoelus, 33
    characteristics of, 36
Proctor Lake, fossils at, 56
Protostega, 48, 51
Pteranodon, characteristics of, 44
pterosaurs, 44-45

Q

Quetzalcoatlus, characteristics of, 45

R

radiometric dating of fossils, 15
reptile-hipped dinosaurs. See *saurischians*.
reptiles
    anapsids, 22
    diapsids, 23
    euryapsids, 23
    relationship of dinosaurs to, 22-23
    synapsids, 23
Rhamphorhynchus, characteristics of, 45

S

San Antonio, Texas, 51
saurischians
    characteristics of, 24, 25
    Acrocanthosaurus, 30
    Alamosaurus, 36
    Coelphysis, 28
    Deinonychus, 31-32
    Ornithomimus, 39
    Pleurocoelus, 33
    Tyrannosaurus, 43
sauropods
    characteristics of, 24, 25
    Alamosaurus, 36
    Diplodocus, 51
    Pleurocoelus, 33
spinosaurids, 30
Stegoceras, characteristics of, 41
stegosaurs, characteristics of, 25
Strecker Museum, Baylor University, 51
synapsids (group of reptiles), 23

T

Tarleton State University, 56
Technosaurus, characteristics of, 29
Tenontosaurus, 56
    characteristics of, 35
Texas Memorial Museum, Austin, Texas, 51
Texas Tech Museum, Lubbock, Texas, 51
Texas Tech University, *Technosaurus* named after, 29
theropods
    characteristics of, 24, 25
    Acrocanthosaurus, 30
    Coelophysis, 28
    Deinonychus, 31-32
    Ornithomimus, 39
    Tyrannosaurus, 43
time scale of existence
    of dinosaurs, 16, 18-20
    "Earth Clock" for showing, 16-18
    of humans, 16
titanosaurs, 36
Torosaurus, 51
    characteristics of, 42
tracks of dinosaurs
    at Glen Rose, 55
    locations of, in Texas, 26
    near Bandera County, 32
    pictures of, 12, 26
    preservation of, 49-50
    site of, 11, 13
    types of, 26
Triassic period, 19, 20
Triceratops, 51
tyrannosaurids, 43
Tyrannosaurus
    characteristics of, 43
    meaning of the word, 15

W

Waco, Texas, 48, 51
    fossils found near, 48
West Texas dinosaurs. See North Central and West Texas dinosaurs.
Witte Museum, San Antonio, Texas, 51

X

Xiphactinus, 56